T0192817

LONDON MATHEMATICAL SOCIETY LECTURE NOTE SERIES

Managing Editor:
Professor M. Reid, Mathematics Institute, University of Warwick, Coventry CV4 7AL, United Kingdom

The titles below are available from booksellers, or from Cambridge University Press at
http://www.cambridge.org/mathematics

London Mathematical Society Lecture Note Series: 410

Representation Theory and Harmonic Analysis of Wreath Products of Finite Groups

TULLIO CECCHERINI-SILBERSTEIN
Università del Sannio, Italy

FABIO SCARABOTTI
Sapienza Università di Roma, Italy

FILIPPO TOLLI
Università Roma Tre, Italy

CAMBRIDGE
UNIVERSITY PRESS

University Printing House, Cambridge CB2 8BS, United Kingdom

Cambridge University Press is part of the University of Cambridge.

It furthers the University's mission by disseminating knowledge in the pursuit of education, learning and research at the highest international levels of excellence.

www.cambridge.org
Information on this title: www.cambridge.org/9781107627857

© T. Ceccherini-Silberstein, F. Scarabotti and F. Tolli 2014

First published 2014

A catalogue record for this publication is available from the British Library

Library of Congress Cataloguing in Publication data
Ceccherini-Silberstein, Tullio.
Representation theory and harmonic analysis of wreath products of finite groups / Tullio Ceccherini-Silberstein, Fabio Scarabotti, and Filippo Tolli.
pages cm
Includes bibliographical references and index.
ISBN 978-1-107-62785-7 (Paperback)
1. Harmonic analysis. 2. Finite groups.
I. Scarabotti, Fabio. II. Tolli, Filippo, 1968– III. Title.
QA403.C44 2014
512´.23–dc23   2013024946

ISBN 978-1-107-62785-7 Paperback

To PVS
To my parents, Cristina, Nadiya, and Virginia
To the memory of my father, and to my mother

Contents

Preface

The aim of these lecture notes is to present an introduction to the representation theory of wreath products of finite groups and to harmonic analysis on the corresponding homogeneous spaces.

The exposition is completely self-contained. The only requirements are the fundamentals of the representation theory of finite groups, for which we refer the possibly inexperienced reader to the monographs by Serre [67], Simon [68], Sternberg [73] and to our recent books [11, 15].

The first chapter constitutes an introduction to the theory of induced representations. It focuses on two main topics, namely harmonic analysis on homogeneous spaces which decompose with multiplicity, and Clifford theory. The latter is developed with the aim of presenting a general formulation of the little group method. The exposition is based on our papers [12, 13, 64].

The second chapter is the core of the monograph. We develop the representation theory of wreath products of finite groups following, in part, the approach by James and Kerber [38] and Huppert [35] and developing our research expository paper [14]. Our approach is both analytical and geometrical. In particular, we interpret the exponentiation and composition actions in terms of actions on suitable finite rooted trees and describe the group of automorphisms of a finite rooted tree as the iterated wreath product of symmetric groups.

We explicitly describe the conjugacy classes of wreath products and the corresponding parameterization of irreducible representations. This is illustrated by a wealth of examples including finite lamplighter groups and the wreath product $S_m \wr S_n$ of symmetric groups.

The third chapter presents an exposition of our recent papers [9, 63, 64] where we develop harmonic analysis on some homogeneous spaces obtained by the composition and the exponentiation actions and their generalization, namely the wreath product of permutation representations, introduced in [64]. As a particular case of the wreath product, we analyze in detail the lamplighter group and develop an exhaustive harmonic analysis on the corresponding finite lamplighter spaces. We also devote a section to the generalized Johnson scheme and to a general construction of finite Gelfand pairs, which we introduced in [9], based on the action of the group of automorphisms of a finite rooted tree on the space of its rooted subtrees.

We wish to express our deepest gratitude to Roger Astley and Gaia Poggiogalli at Cambridge University Press and to Susan Parkinson, our copy-editor, for their most kind, constant and valuable help at all stages of the editing process.

Roma, April 2013   TCS, FS and FT

1

General theory

In this chapter we discuss the notion of an induced representation and the structure of the commutant of a representation, and we present a new approach to Clifford theory. We assume the reader to be familiar with the basic rudiments of the representation theory of finite groups. We refer to the monographs by Bump [7], Fulton and Harris [29], Isaacs [36], Serre [67], Simon [68] and Sternberg [73] as basic references; see also our monograph [15].

In Section 1.1 we present the main properties of induction, focusing on the Frobenius character formula and Frobenius reciprocity. Then, in Section 1.2, we discuss several aspects of Frobenius reciprocity for a permutation representation; in particular, we show that the spherical Fourier transform provides an explicit isomorphism between the commutant of a permutation representation and the algebra of bi-K-invariant functions. In the last part of the section we examine the particular case of a multiplicity free permutation representation, which yields the notion of a Gelfand pair. Finally, in Section 1.3, we present an exposition of Clifford theory, which provides a powerful tool for relating the representation theory of a group G and the representation theory of a normal subgroup $N \leq G$.

1.1 Induced representations

The presentation of this section was inspired by the books by Bump [7], Serre [67] and Sternberg [73] and by our research-expository paper [1?].

1.1.1 Definitions

Let G be a finite group. Let K be a subgroup of G and (ρ, W) a representation of K. Let $S \subseteq G$ be a system of representatives for the left cosets of K in G, so that

$$G = \coprod_{s \in S} sK \tag{1.1}$$

(here, and in what follows, $\coprod$ denotes a disjoint union). We shall always suppose that the unit element 1_G of G belongs to S. Consider the space $W^S = \{f : S \to W\}$. Given $g \in G$, we define the linear map $\sigma(g) : W^S \to W^S$ by setting

$$[\sigma(g)f](s) = \rho(k^{-1})[f(t)] \tag{1.2}$$

for all $f \in W^S$ and $s \in S$, where $t \in S$ and $k \in K$ are the unique elements such that $g^{-1}s = tk$.

Let $g_1, g_2 \in G$, $f \in W^S$ and $s \in S$. Also let $s_1, s_2 \in S$ and $k_1, k_2 \in K$ such that $g_1^{-1}s = s_1k_1$ and $g_2^{-1}s_1 = s_2k_2$. Note that $(g_1g_2)^{-1}s = s_2(k_2k_1)$. We then have

$$\begin{aligned}
[\sigma(g_1)(\sigma(g_2)f)](s) &= \rho(k_1^{-1})[(\sigma(g_2)f)(s_1)] \\
&= \rho(k_1^{-1})\rho(k_2^{-1})[f(s_2)] \\
&= \rho(k_2k_1)^{-1}[f(s_2)] \\
&= [\sigma(g_1g_2)f](s).
\end{aligned}$$

It follows that (σ, W^S) is a representation of G.

We will always suppose that the representations (ρ, W) are *unitary*, that is, W is equipped with a scalar product $\langle \cdot, \cdot \rangle_W$ and one has

$$\langle \rho(g)w, \rho(g)w' \rangle_W = \langle w, w' \rangle_W$$

for all $g \in G$ and $w, w' \in W$.

Remark 1.1.1   Note that if (ρ, W) is unitary then (σ, W^S) is also unitary with respect to the scalar product on W^S defined by

$$\langle f, f' \rangle_{W^S} = \sum_{s \in S} \langle f(s), f'(s) \rangle_W$$

for all $f, f' \in W^S$. Indeed, for all $f, f' \in W^S$ and $g \in G$ we have

$$\begin{aligned}
\langle \sigma(g)f, \sigma(g)f' \rangle_{W^S} &= \sum_{s \in S} \langle [\sigma(g)f](s), [\sigma(g)f'](s) \rangle_W \\
&= \sum_{s \in S} \langle \rho(k_s^{-1})f(t_s), \rho(k_s^{-1})f'(t_s) \rangle_W
\end{aligned}$$

$$= \sum_{s \in S} \langle f(t_s), f'(t_s) \rangle w$$

$$= \sum_{t \in S} \langle f(t), f'(t) \rangle w$$

$$= \langle f, f' \rangle_{W^S}$$

where, for all $s \in S$, $t_s \in S$ and $k_s \in K$ are the unique elements such that $g^{-1}s = t_s k_s$. Note that, since the map $sK \mapsto t_s K$, that is, left multiplication by g^{-1}, yields a bijection on the left cosets of K in G, then the map $s \mapsto t_s$ is a bijection on S.

Consider now the space $W^G = \{f : G \to W\}$ and define the subspace $Z \subseteq W^G$ by setting

$$Z = \{f \in W^G : f(gk) = \rho(k^{-1})f(g) \text{ for all } g \in G \text{ and } k \in K\}. \quad (1.3)$$

Given $g_1 \in G$ define the linear map $\theta(g_1) : Z \to Z$ by setting

$$[\theta(g_1)f](g_2) = f(g_1^{-1}g_2) \qquad (1.4)$$

for all $g_2 \in G$ and $f \in Z$. Let $g_1, g_2, g_3 \in G$, $k \in K$ and $f \in Z$. We then have

$$[\theta(g_1)f](g_2 k) = f(g_1^{-1}g_2 k) = \rho(k^{-1})f(g_1^{-1}g_2) = \rho(k^{-1})[\theta(g_1)f](g_2),$$

which shows that $\theta(g_1)f \in Z$. Moreover,

$$[\theta(g_1 g_2)f](g_3) = f((g_1 g_2)^{-1}g_3)$$
$$= f(g_2^{-1}g_1^{-1}g_3)$$
$$= [\theta(g_2)f](g_1^{-1}g_3)$$
$$= [\theta(g_1)\theta(g_2)f](g_3).$$

This shows that (θ, Z) is a representation of G.

Proposition 1.1.2 *The G-representations (θ, Z) and (σ, W^S) are equivalent. In particular, the equivalence class of the representation (σ, W^S) does not depend on the particular choice of the set of representatives for the left cosets of the subgroup.*

Proof   Consider the map $\Phi : W^S \to W^G$ defined by setting

$$\Phi(f)(g) = \rho(k^{-1})f(s)$$

for all $f \in W^S$ and $g \in G$, where $s \in S$ and $k \in K$ are the unique elements such that $g = sk$.

Let $g \in G$ and $k \in K$. Let also $s \in S$ and $h \in K$ such that $g = sh$ and note that $gk = s(hk)$. We have

$$\begin{aligned}
\Phi(f)(gk) &= \rho((hk)^{-1})f(s) \\
&= \rho(k^{-1})\rho(h^{-1})f(s) \\
&= \rho(k^{-1})\Phi(f)(g).
\end{aligned}$$

This shows that the image of Φ is contained in Z (see (1.3)).

Note that Φ is a bijection of W^S onto Z, since every element in Z is uniquely determined by its restriction to S.

Let us show that Φ intertwines the representations σ and θ. Let $g_1, g_2 \in G$ and $f \in W^S$. Let also $s_2, s \in S$ and $k_2, k \in K$ such that $g_2 = s_2 k_2$ and $g_1^{-1}s_2 = sk$ and observe that $skk_2 = g_1^{-1}s_2k_2 = g_1^{-1}g_2$. Then we have

$$\begin{aligned}
\Phi[\sigma(g_1)f](g_2) &= \rho(k_2^{-1})[\sigma(g_1)f(s_2)] \\
&= \rho(k_2^{-1})[\rho(k^{-1})f(s)] \\
&= \rho((kk_2)^{-1})f(s) \\
&= \Phi(f)(skk_2) \\
&= \Phi(f)(g_1^{-1}g_2) \\
&= [\theta(g_1)\Phi(f)](g_2).
\end{aligned}$$

This shows that Φ intertwines the representations σ and θ. It follows that $\sigma \sim \theta$. $\square$

Definition 1.1.3   The G-representation (σ, W^S), where σ is defined by (1.2), is called the *representation induced* by (ρ, W) from K to G and is denoted by $(\mathrm{Ind}_K^G \rho, \mathrm{Ind}_K^G W)$.

It follows from Proposition 1.1.2 that the induced representation is, up to equivalence, independent of the choice of the system S of representatives for the left cosets of K in G. Recalling that the *dimension* of a representation (ρ, W) is defined as $\dim(W)$, we have that the dimension of the induced representation $(\mathrm{Ind}_K^G \rho, \mathrm{Ind}_K^G W)$ is given by

$$\dim(\mathrm{Ind}_K^G W) = [G : K]\dim(W), \tag{1.5}$$

which immediately follows from the equalities $\dim(W^S) = |S|\dim(W)$ and $|S| = [G : K]$.

Moreover, from (1.2) we deduce that

$$[\mathrm{Ind}_K^G \rho(g)f](s) = \rho(s^{-1}gt)[f(t)] \tag{1.6}$$

where t is the unique element in S such that $s^{-1}gt \in K$. As a consequence, setting $\rho(g') = 0$ if $g' \in G \setminus K$, we can represent the linear map $\mathrm{Ind}_K^G \rho(g)$ by the $S \times S$ matrix with entries in $\mathrm{End}(W)$ given by

$$\left(\rho(s_1^{-1}gs_2)\right)_{s_1,s_2 \in S}. \tag{1.7}$$

Exercise 1.1.4   Show that, for $g_1, g_2 \in G$, the matrix representing the linear map $\mathrm{Ind}_K^G \rho(g_1g_2)$ is the product of the matrices representing $\mathrm{Ind}_K^G \rho(g_1)$ and $\mathrm{Ind}_K^G \rho(g_2)$.

Given a finite set X we denote by $\mathrm{Sym}(X)$ the *symmetric group on X*, that is, the set of all bijective maps (called *permutations*) $\pi : X \to X$ with multiplication given by composition. When $|X| = n$, we denote $\mathrm{Sym}(X)$ by S_n and refer to it as to the *symmetric group of degree n*.

Example 1.1.5   Let $G = S_3 = \{e, (12), (13), (23), (123), (132)\}$. Consider the subgroup $K = \{e, (12)\} \cong S_2$. We choose as a set of representatives for the left cosets of K in G the set $S = \{e, (123), (132)\}$. Note that $(13) = (123)(12)$ and $(23) = (132)(12)$. The unique representations of K are one dimensional, namely the *trivial representation* $(\iota, \mathbb{C})$ and the *alternating representation* $(\varepsilon, \mathbb{C})$. For simplicity of notation, we denote by $\overline{\iota}$ and $\overline{\varepsilon}$ the corresponding induced representations of G on $\mathbb{C} \oplus (123)\mathbb{C} \oplus (132)\mathbb{C}$ (note that here $W = \mathbb{C}$). Given $f \in \mathbb{C} \oplus (123)\mathbb{C} \oplus (132)\mathbb{C}$ we then have

$[\overline{\iota}(e)f](e) = f(e),$	$[\overline{\iota}(e)f](123) = f(123),$	$[\overline{\iota}(e)f](132) = f(132);$
$[\overline{\iota}(12)f](e) = f(e),$	$[\overline{\iota}(12)f](123) = f(132),$	$[\overline{\iota}(12)f](132) = f(123);$
$[\overline{\iota}(13)f](e) = f(123),$	$[\overline{\iota}(13)f](123) = f(e),$	$[\overline{\iota}(13)f](132) = f(132);$
$[\overline{\iota}(23)f](e) = f(132),$	$[\overline{\iota}(23)f](123) = f(123),$	$[\overline{\iota}(23)f](132) = f(e);$
$[\overline{\iota}(123)f](e) = f(132),$	$[\overline{\iota}(123)f](123) = f(e),$	$[\overline{\iota}(123)f](132) = f(123);$
$[\overline{\iota}(132)f](e) = f(123),$	$[\overline{\iota}(132)f](123) = f(132),$	$[\overline{\iota}(132)f](132) = f(e)$

The corresponding matrices as in (1.7) are given by

$$\overline{\iota}(e) = \begin{pmatrix} 1 & 0 & 0 \\ 0 & 1 & 0 \\ 0 & 0 & 1 \end{pmatrix}, \quad \overline{\iota}(12) = \begin{pmatrix} 1 & 0 & 0 \\ 0 & 0 & 1 \\ 0 & 1 & 0 \end{pmatrix}, \quad \overline{\iota}(13) = \begin{pmatrix} 0 & 1 & 0 \\ 1 & 0 & 0 \\ 0 & 0 & 1 \end{pmatrix},$$

$$\bar{\iota}(23) = \begin{pmatrix} 0 & 0 & 1 \\ 0 & 1 & 0 \\ 1 & 0 & 0 \end{pmatrix}, \quad \bar{\iota}(123) = \begin{pmatrix} 0 & 0 & 1 \\ 1 & 0 & 0 \\ 0 & 1 & 0 \end{pmatrix}, \quad \bar{\iota}(132) = \begin{pmatrix} 0 & 1 & 0 \\ 0 & 0 & 1 \\ 1 & 0 & 0 \end{pmatrix}.$$

Similarly,

$$[\bar{\varepsilon}(e)f](e) = f(e), \qquad [\bar{\varepsilon}(e)f](123) = f(123), \qquad [\bar{\varepsilon}(e)f](132) = f(132);$$
$$[\bar{\varepsilon}(12)f](e) = -f(e), \qquad [\bar{\varepsilon}(12)f](123) = -f(132), \qquad [\bar{\varepsilon}(12)f](132) = -f(123);$$
$$[\bar{\varepsilon}(13)f](e) = -f(123), \qquad [\bar{\varepsilon}(13)f](123) = -f(e), \qquad [\bar{\varepsilon}(13)f](132) = -f(132);$$
$$[\bar{\varepsilon}(23)f](e) = -f(132), \qquad [\bar{\varepsilon}(23)f](123) = -f(123), \qquad [\bar{\varepsilon}(23)f](132) = -f(e);$$
$$[\bar{\varepsilon}(123)f](e) = f(132), \qquad [\bar{\varepsilon}(123)f](123) = f(e), \qquad [\bar{\varepsilon}(123)f](132) = f(123);$$
$$[\bar{\varepsilon}(132)f](e) = f(123), \qquad [\bar{\varepsilon}(132)f](123) = f(132), \qquad [\bar{\varepsilon}(132)f](132) = f(e),$$

and the corresponding matrices, again as in (1.7), are given by

$$\bar{\varepsilon}(e) = \begin{pmatrix} 1 & 0 & 0 \\ 0 & 1 & 0 \\ 0 & 0 & 1 \end{pmatrix}, \quad \bar{\varepsilon}(12) = \begin{pmatrix} -1 & 0 & 0 \\ 0 & 0 & -1 \\ 0 & -1 & 0 \end{pmatrix}, \quad \bar{\varepsilon}(13) = \begin{pmatrix} 0 & -1 & 0 \\ -1 & 0 & 0 \\ 0 & 0 & -1 \end{pmatrix}$$

$$\bar{\varepsilon}(23) = \begin{pmatrix} 0 & 0 & -1 \\ 0 & -1 & 0 \\ -1 & 0 & 0 \end{pmatrix}, \quad \bar{\varepsilon}(123) = \begin{pmatrix} 0 & 0 & 1 \\ 1 & 0 & 0 \\ 0 & 1 & 0 \end{pmatrix}, \quad \bar{\varepsilon}(132) = \begin{pmatrix} 0 & 1 & 0 \\ 0 & 0 & 1 \\ 1 & 0 & 0 \end{pmatrix}.$$

Definition 1.1.6 Let G be a finite group acting transitively on a finite set X and denote by $L(X)$ the space of all functions $f : X \to \mathbb{C}$. Then the corresponding *permutation representation* $(\lambda, L(X))$ is defined by

$$[\lambda(g)f](x) = f(g^{-1}x)$$

for all $g \in G$, $f \in L(X)$, and $x \in X$.

Fix a point $x_0 \in X$ and denote by $K = \{g \in G : gx_0 = x_0\}$ its *stabilizer*; then we can identify X and the set G/K of left cosets of K as G-spaces. We refer to $X = G/K$ as an *homogeneous space*. Denote by $(\iota, \mathbb{C})$ the trivial (one-dimensional) representation of K.

Proposition 1.1.7 *The permutation representation λ and the induced representation* $\mathrm{Ind}_K^G \iota$ *are equivalent.*

Proof Let $\Phi : L(X) \to \mathbb{C}^S \equiv \mathrm{Ind}_K^G \mathbb{C}$ be the map defined by

$$[\Phi f](s) = f(sx_0)$$

for all $f \in L(X)$ and $s \in S$. Clearly Φ is a vector space isomorphism. Moreover, for all $f \in L(X)$, $g \in G$, and $s \in S$, if $t \in S$ and $k \in K$ are such that $g^{-1}s = tk$ then we have

$$[\Phi(\lambda(g)f)](s) = [\lambda(g)f](sx_0)$$
$$= f(g^{-1}sx_0)$$
$$= f(tx_0)$$
$$= [\Phi f](t)$$
$$= [\text{Ind}_K^G \iota(g)\Phi f](s).$$

This shows that Φ intertwines λ and $\text{Ind}_K^G \iota$. It is easy to see that Φ is a bijection and therefore $\lambda \sim \text{Ind}_K^G \iota$. $\qquad\qquad\qquad\qquad\qquad\qquad\qquad\qquad\qquad\qquad\square$

Definition 1.1.8   Suppose that G is a finite group, K a subgroup of G, and (σ, V) a representation of G. The *restriction* $\text{Res}_K^G \sigma$ of σ from G to K is the representation of K on V defined by setting $\text{Res}_K^G \sigma(k) = \sigma(k)$ for all $k \in K$.

In the notation of Definition 1.1.8, if $W \leq V$ is a *K-invariant* subspace (that is, $\sigma(k)w \in W$ for all $k \in K$ and $w \in W$) then we may also consider the representation $\text{Res}_K^G \sigma|_W$ of K on W (here, and in what follows, given sets X, Y and $Z \subseteq X$ and a map $f : X \to Y$ we denote by $f|_Z : Z \to Y$ the *restriction map* defined by $f|_Z(x) = f(x)$ for all $x \in Z$).

In the notation of Definition 1.1.3, set $W_0 = \{f \in W^S : f(s) = 0 \text{ if } s \neq 1_G\}$. Then we have

(i) $\text{Res}_K^G \sigma|_{W_0}$ is equivalent to ρ;
(ii) $\text{Ind}_K^G W = \bigoplus_{s \in S} \sigma(s)W_0$.

Indeed (i) is obvious, while to prove (ii) it suffices to note that for $s, t \in S$ and $f_0 \in W_0$ we have

$$[\sigma(s)f_0](t) = \rho(k^{-1})f_0(q) = \begin{cases} 0 & \text{if } s \neq t \\ f_0(1_G) & \text{if } s = t \end{cases}$$

where $q \in S, k \in K$ are chosen in such a way that $s^{-1}t = qk$. These two properties provide a characterization of induced representations (or an alternative definition; see [67]):

Proposition 1.1.9   *Let G, K, and S be as above. Let (σ, V) be a representation of G. Let $W \leq V$ be a K-invariant subspace and set $\rho = \text{Res}_K^G \sigma|_W$. Suppose that*

$$V = \bigoplus_{s \in S} \sigma(s)W.$$

Then $\text{Ind}_K^G \rho \sim \sigma$.

Proof First let us set $\tau = \operatorname{Ind}_K^G \rho$. Consider the linear map

$$\Phi : W^S \longrightarrow V$$
$$f \longmapsto \bigoplus_{s \in S} \sigma(s) f(s).$$

It is clearly a bijection. Moreover, we have, for $f \in W^S$, $g \in G$ and $s \in S$,

$$[\tau(g)f](s) = \rho(k_s^{-1}) f(t_s) \equiv \sigma(k_s^{-1}) f(t_s),$$

where $t_s \in S$ and $k_s \in K$ are chosen in such a way that $g^{-1}s = t_s k_s$. Therefore

$$\begin{aligned}
\Phi\tau(g)f &= \bigoplus_{s \in S} \sigma(s)[\tau(g)f](s) \\
&= \bigoplus_{s \in S} \sigma(s)\sigma(k_s^{-1}) f(t_s) \\
&= \bigoplus_{s \in S} \sigma(g)\sigma(t_s) f(t_s) \qquad (sk_s^{-1} = gt_s) \\
&= \sigma(g) \bigoplus_{s \in S} \sigma(t_s) f(t_s) \\
&= \sigma(g)\Phi f.
\end{aligned}$$

This proves that Φ also intertwines $\operatorname{Ind}_K^G \rho$ and σ. It follows that the two representations are equivalent.   □

1.1.2 Transitivity and additivity of induction

One of the most important properties of induction is transitivity.

Proposition 1.1.10 (Induction in stages) *Let G be a finite group, $K \leq H \leq G$ be subgroups and (ρ, W) a representation of K. Then*

$$\operatorname{Ind}_H^G(\operatorname{Ind}_K^H W) \cong \operatorname{Ind}_K^G W \tag{1.8}$$

as G-representations.

Proof Let $T \subset H$ (resp. $S \subset G$) be a set of representatives for the left cosets of K in H (resp. of H in G). Then the map $(s, t) \to st$ establishes a bijection between $S \times T$ and the set $ST = \{st : s \in S, t \in T\} \subset G$. Moreover, ST is a set of representatives for the left cosets of K in G.

Consider the map

$$\Phi : W^{ST} \to (W^T)^S \tag{1.9}$$

defined by $[\Phi f(s)](t) = f(st)$ for all $f \in W^{ST}$, $s \in S$ and $t \in T$. It follows from the above observations that Φ is a linear isomorphism. Let us show that Φ intertwines the representations $\mathrm{Ind}_K^G \rho$ and $\mathrm{Ind}_H^G \mathrm{Ind}_K^H \rho$. Let $f \in W^{ST}$, $g \in G$, $s \in S$ and $t \in T$. Denote by $s' \in S$, $t' \in T$ and $k \in K$ the unique elements such that $g^{-1}(st) = s't'k$. Also set

$$h = t'kt^{-1} \in H. \tag{1.10}$$

We then have on the one hand

$$[\mathrm{Ind}_K^G \rho(g) f](st) = \rho(k^{-1}) f(s't')$$

so that

$$\{\Phi[\mathrm{Ind}_K^G \rho(g) f](s)\}(t) = \rho(k^{-1})\left\{[\Phi f(s')](t')\right\}. \tag{1.11}$$

On the other hand we have, using (1.10) (that is, the identities $g^{-1}s = s'h$ and $(h^{-1})^{-1}t = ht = t'k$),

$$\left\{[\mathrm{Ind}_H^G \mathrm{Ind}_K^H \rho(g)(\Phi f)](s)\right\}(t) = \left\{\mathrm{Ind}_K^H \rho(h^{-1})[\Phi f(s')]\right\}(t)$$

$$= \rho(k^{-1})\left\{[\Phi f(s')](t')\right\}. \tag{1.12}$$

Comparing (1.11) and (1.12), the proof is complete.   □

Another property of the induction operation is additivity.

Proposition 1.1.11   *Let G be a finite group and $K \leq G$ a subgroup. Let (ρ_1, W_1) and (ρ_2, W_2) be two representations of K. Then*

$$\mathrm{Ind}_K^G(\rho_1 \oplus \rho_2) \sim \mathrm{Ind}_K^G(\rho_1) \oplus \mathrm{Ind}_K^G(\rho_2). \tag{1.13}$$

Proof   Let $S \subset G$ be a set of representatives for the left cosets of K in G. Consider the linear map $\Phi: (W_1 \oplus W_2)^S \to W_1^S \oplus W_2^S$ dehned by

$$[\Phi(f)]_1(s) = [f(s)]_1 \qquad \text{and} \qquad [\Phi(f)]_2(s) = [f(s)]_2$$

for all $f \in (W_1 \oplus W_2)^S$ and $s \in S$. It it clear that Φ is a linear isomorphism. Let us show that Φ intertwines $\mathrm{Ind}_K^G(\rho_1 \oplus \rho_2)$ and $\mathrm{Ind}_K^G(\rho_1) \oplus \mathrm{Ind}_K^G(\rho_2)$. Let $g \in G$,

$s \in S$ and $f = (W_1 \oplus W_2)^S$. Then we can find unique $t \in S$ and $k \in K$ such that $g^{-1}s = tk$. We then have, for $i = 1, 2$,

$$
\begin{aligned}
[\Phi(\operatorname{Ind}_K^G(\rho_1 \oplus \rho_2)(g)f)]_i(s) &= [(\operatorname{Ind}_K^G(\rho_1 \oplus \rho_2)(g)f)(s)]_i \\
&= [(\rho_1(k^{-1}) \oplus \rho_2(k^{-1})f(t)]_i \\
&= \rho_i(k^{-1})[f(t)]_i \\
&= \rho_i(k^{-1})[\Phi(f)]_i(t) \\
&= \operatorname{Ind}_K^G(\rho_i)(g)[\Phi(f)]_i(s) \\
&= [(\operatorname{Ind}_K^G(\rho_1) \oplus \operatorname{Ind}_K^G(\rho_2))(g)\Phi(f)]_i(s).
\end{aligned}
$$

This shows that Φ intertwines $\operatorname{Ind}_K^G(\rho_1 \oplus \rho_2)$ and $\operatorname{Ind}_K^G(\rho_1) \oplus \operatorname{Ind}_K^G(\rho_2)$ and the proof is complete.   □

1.1.3 Frobenius character formula

Theorem 1.1.12 (Frobenius character formula for induced representations)
Let G be a finite group, $K \leq G$ a subgroup and $S \subseteq G$ a system of representatives of left cosets of K in G. Let (ρ, W) be a representation of K and denote by χ_ρ its character. Then the character $\chi_{\operatorname{Ind}_K^G \rho}$ of the induced representation is given by

$$
\chi_{\operatorname{Ind}_K^G \rho}(g) = \sum_{\substack{s \in S: \\ s^{-1}gs \in K}} \chi_\rho(s^{-1}gs). \tag{1.14}
$$

Proof Let $\langle \cdot, \cdot \rangle_W$ denote the scalar product on W and let $\{e_j : j \in J\}$ be an orthonormal basis for W. Then, we define a scalar product on W^S as in Remark 1.1.1. Also, denote by $f_{s,j} \in W^S$ the map defined by $f_{s,j}(s') = \delta_{s,s'}e_j \in W$ for all $s, s' \in S$ and $j \in J$. It follows that $\{f_{s,j} : s \in S, j \in J\}$ is an orthonormal basis for $W^S = \operatorname{Ind}_K^G W$.

Now let $g \in G$ and $s \in S$. Then there exist unique $t \in S$ and $k \in K$ such that $g^{-1}s = tk$ or, equivalently,

$$
k^{-1} = s^{-1}gt. \tag{1.15}
$$

Setting $\sigma = \operatorname{Ind}_K^G \rho$ we deduce that

$$
[\sigma(g)f_{u,j}](s) = \rho(k^{-1})f_{u,j}(t) = \delta_{u,t}\rho(k^{-1})e_j \tag{1.16}
$$

for all $u \in S$ and $j \in J$. We then have

$$
\chi_{\mathrm{Ind}_K^G \rho}(g) = \sum_{\substack{u \in S \\ j \in J}} \langle \sigma(g) f_{u,j}, f_{u,j} \rangle_{WS} = \sum_{\substack{u,s \in S \\ j \in J}} \langle [\sigma(g) f_{u,j}](s), f_{u,j}(s) \rangle_W
$$

$$
= \sum_{\substack{u,s \in S \\ j \in J}} \delta_{u,t} \delta_{u,s} \langle \rho(k^{-1}) e_j, e_j \rangle_W \qquad \text{(by (1.16))}
$$

$$
= \sum_{\substack{s \in S \\ j \in J}} \delta_{s,t} \langle \rho(k^{-1}) e_j, e_j \rangle_W
$$

$$
= \sum_{\substack{s \in S: \\ s^{-1} g s \in K \\ j \in J}} \langle \rho(s^{-1} g s) e_j, e_j \rangle_W \qquad \text{(by (1.15))}
$$

$$
= \sum_{\substack{s \in S: \\ s^{-1} g s \in K}} \chi_\rho(s^{-1} g s). \qquad \square
$$

Exercise 1.1.13 Give an alternative proof of the Frobenius character formula (1.14) by using the matrix representation (1.7).

1.1.4 Induction and restriction

Induction and restriction are not inverse operations: this follows immediately after comparing dimensions (see (1.5)). The following example shows that, even in the case of the trivial representation, the composition of these two operations is quite different from the identity.

Example 1.1.14 Let K be a subgroup of a group G and set $X = G/K$. Consider first the trivial representation $(\iota_G, \mathbb{C})$ of G. We clearly have $\mathrm{Res}_K^G \iota_G = \iota_K$. Therefore $\mathrm{Ind}_K^G \mathrm{Res}_K^G \iota_G = \mathrm{Ind}_K^G \iota_K = (\lambda, L(X))$ (cf. Definition 1.1.6 and Proposition 1.1.7). Thus, if $K \neq G$ we easily deduce that $\mathrm{Ind}_K^G \mathrm{Res}_K^G \iota_G \neq \tilde{\iota}_G$, where $(\tilde{\iota}_G, L(X))$ is the G-representation defined by $\tilde{\iota}_G(g) f = f$ for all $g \in G$ and $f \in L(X)$.

Consider now the trivial representation $(\iota_K, \mathbb{C})$ of K. Let us show that if K is not a normal subgroup then $\mathrm{Res}_K^G \mathrm{Ind}_K^G \iota_K \neq \tilde{\iota}_K$, where $(\tilde{\iota}_K, L(X))$ is the K-representation defined by $\tilde{\iota}_K(k) f = f$ for all $k \in K$ and $f \in L(X)$. Denoting by $x_0 \in X$ a point fixed by K, by our assumption we can find $k \in K$ and

$g \in G$ such that $g^{-1}k^{-1}g \notin K$ or equivalently $k^{-1}gx_0 \neq gx_0$. Then, recalling that $\operatorname{Ind}_K^G \iota_K = (\lambda, L(X))$, we have

$$[\operatorname{Res}_K^G \operatorname{Ind}_K^G \iota_K(k)\delta_{gx_0}](gx_0) = [\operatorname{Res}_K^G \lambda(k)\delta_{gx_0}](gx_0)$$
$$= [\lambda(k)\delta_{gx_0}](gx_0)$$
$$= \delta_{gx_0}(k^{-1}gx_0)$$
$$= 0 \neq 1$$
$$= [\delta_{gx_0}](gx_0)$$
$$= [\tilde{\iota}_K(k)\delta_{gx_0}](gx_0),$$

where $\delta_x \in L(X)$ is the *Dirac function* at $x \in X$ (that is, $\delta_x(y) = 1$ if $x = y$ and 0 otherwise).

However, the following two results establish the right connection between induction and restriction.

Proposition 1.1.15 *Let (θ, V) be a representation of a group G and (ρ, W) a representation of a subgroup $K \leq G$. Then we have the following isomorphism of G-representations:*

$$V \otimes \operatorname{Ind}_K^G W \cong \operatorname{Ind}_K^G[(\operatorname{Res}_K^G V) \otimes W]. \tag{1.17}$$

Proof Let $S \subset G$ be a set of representatives for the left cosets of K in G. Consider the linear map $\Phi \colon V \otimes W^S \to (V \otimes W)^S$ defined on simple tensors by

$$[\Phi(v \otimes f)](s) = \theta(s^{-1})v \otimes f(s)$$

for all $v \in V$, $f \in W^S$ and $s \in S$. Let us prove that Φ is a linear isomorphism.

To show that it is surjective, fix $F \in \operatorname{Ind}_K^G[(\operatorname{Res}_K^G V) \otimes W]$. Then, for every $s \in S$ there exists a finite index set I_s and $v_i^s \in V$, $w_i^s \in W$, $i \in I_s$, such that $F(s) = \sum_{i \in I_s} v_i^s \otimes w_i^s$. Consider the element $\pi = \sum_{\substack{i \in I_t \\ t \in S}} \theta(t)v_i^t \otimes f_i^t \in V \otimes \operatorname{Ind}_K^G W$ where $f_i^t(s) = \delta_{s,t}w_i^s$ for all $t \in S$. We claim that $\Phi(\pi) = F$. Indeed we have, for all $s \in S$,

$$[\Phi(\pi)](s) = \sum_{\substack{i \in I_t \\ t \in S}} \theta(s^{-1})\theta(t)v_i^t \otimes f_i^t(s)$$
$$= \sum_{i \in I_s} v_i^s \otimes w_i^s$$
$$= F(s).$$

This shows that Φ is surjective. Moreover, using (1.5) we have

$$\dim(V \otimes \text{Ind}_K^G W) = \dim(V)\dim(W)[G:K] = \dim(\text{Ind}_K^G[(\text{Res}_K^G V) \otimes W]),$$

so that, by linearity, we deduce that Φ is also injective. Thus Φ is a linear isomorphism between $V \otimes \text{Ind}_K^G W$ and $\text{Ind}_K^G[(\text{Res}_K^G V) \otimes W]$.

Let us show that Φ intertwines $\theta \otimes \text{Ind}_K^G \rho$ and $\text{Ind}_K^G(\text{Res}_K^G \theta \otimes \rho)$. Let $v \in V$, $f \in W^S, g \in G$ and $s \in S$. Then we can find unique elements $t \in T$ and $k \in K$ such that $g^{-1}s = tk$ (so that $s^{-1}g = k^{-1}t^{-1}$). We have on the one hand

$$\left\{ \Phi[(\theta \otimes \text{Ind}_K^G \rho)(g)(v \otimes f)] \right\}(s) = \left\{ \Phi[\theta(g)v \otimes \rho(k^{-1})f(t)] \right\}(s)$$
$$= \theta(s^{-1})\theta(g)v \otimes \rho(k^{-1})f(t)$$
$$= \theta(k^{-1}t^{-1})v \otimes \rho(k^{-1})f(t) \qquad (1.18)$$

and, on the other hand,

$$[\text{Ind}_K^G(\text{Res}_K^G \theta \otimes \rho)(g)\Phi(v \otimes f)](s) = [(\text{Res}_K^G \theta(k^{-1}) \otimes \rho(k^{-1}))\Phi(v \otimes f)](t)$$
$$= [\theta(k^{-1}) \otimes \rho(k^{-1})][\theta(t^{-1})v \otimes f(t)]$$
$$= \theta(k^{-1})\theta(t^{-1})v \otimes \rho(k^{-1})f(t).$$
$$(1.19)$$

A comparison between (1.18) and (1.19) shows that Φ is an intertwiner, completing the proof.   □

Corollary 1.1.16 *Let G be a group and $K \leq G$ a subgroup, denote by X the homogeneous space G/K and let (θ, V) be a representation of G. Then we have*

$$\text{Ind}_K^G \text{Res}_K^G V \cong V \otimes L(X). \qquad (1.20)$$

Proof Apply Proposition 1.1.15 with $(\rho, W) = (\iota_K, \mathbb{C})$, the trivial representation of K. In this case, $\text{Ind}_K^G W = L(X)$ and $(\text{Res}_K^G V) \otimes W = (\text{Res}_K^G V) \otimes \mathbb{C} \cong \text{Res}_K^G V$.   □

Remark 1.1.17 Let G be a group and $K \leq G$ a subgroup. If (σ_1, V_1) and (σ_2, V_2) are two representations of G then we clearly have

$$\text{Res}_K^G(\sigma_1 \otimes \sigma_2) \sim \text{Res}_K^G(\sigma_1) \otimes \text{Res}_K^G(\sigma_2).$$

On the contrary, if K is a proper subgroup of G and (ρ_1, W_1) and (ρ_2, W_2) are two representations of K then we always have

$$\text{Ind}_K^G(\rho_1 \otimes \rho_2) \not\sim \text{Ind}_K^G(\rho_1) \otimes \text{Ind}_K^G(\rho_2),$$

as immediately follows from (1.5).

1.1.5 Induced representations and induced operators

Let G be a finite group and $K \leq G$ a subgroup. Let (ρ_1, W_1) and (ρ_2, W_2) be two representations of K and let $T \colon W_1 \to W_2$ be a K-intertwining operator. Fix a set $S \subset G$ of representatives for the set of left cosets of K in G. We define an operator $\operatorname{Ind}_K^G T \colon \operatorname{Ind}_K^G W_1 \to \operatorname{Ind}_K^G W_2$ by setting

$$[\operatorname{Ind}_K^G T(f)](s) = T[f(s)] \qquad (1.21)$$

for all $f \in V_1 = W_1^S$ and $s \in S$.

Proposition 1.1.18 *With the above notation we have*

(i) $\operatorname{Ind}_K^G T$ *intertwines* $\operatorname{Ind}_K^G \rho_1$ *and* $\operatorname{Ind}_K^G \rho_2$;
(ii) $\operatorname{Ker} \operatorname{Ind}_K^G T = \operatorname{Ind}_K^G \operatorname{Ker} T$;
(iii) $\operatorname{Ran} \operatorname{Ind}_K^G T = \operatorname{Ind}_K^G \operatorname{Ran} T$.

Proof Let $f \in W_1^S$, $g \in G$ and $s \in S$. Then we can find unique $t \in S$ and $k \in K$ such that $g^{-1}s = tk$ and therefore

$$\begin{aligned}
\left\{ \operatorname{Ind}_K^G \rho_2(g)[\operatorname{Ind}_K^G T(f)] \right\}(s) &= \rho_2(k^{-1}) \left\{ [\operatorname{Ind}_K^G T(f)](t) \right\} \\
&= \rho_2(k^{-1}) T[f(t)] \\
&= T[\rho_1(k^{-1}) f(t)] \\
&= T \left\{ [\operatorname{Ind}_K^G \rho_1(g) f](s) \right\} \\
&= \left\{ \operatorname{Ind}_K^G T[\operatorname{Ind}_K^G \rho_1(g) f] \right\}(s).
\end{aligned}$$

This shows (i). To show the remaining part, let $f \in V_1 = W_1^S$.

We have $f \in \operatorname{Ker} \operatorname{Ind}_K^G T$ if and only if $0 = [\operatorname{Ind}_K^G T(f)](s) = T[f(s)]$ for all $s \in S$, that is, if and only if $f(s) \in \operatorname{Ker} T$ for all $s \in S$, equivalently, $f \in \operatorname{Ind}_K^G \operatorname{Ker} T$. This shows (ii).

Similarly, we have $f \in \operatorname{Ran} \operatorname{Ind}_K^G T$ if and only if there exists $f' \in \operatorname{Ind}_K^G W_1 = W_1^S$ such that $f(s) = [\operatorname{Ind}_K^G T(f')](s) = T[f'(s)]$ for all $s \in S$, that is, if and only if $f(s) \in \operatorname{Ran} T$ for all $s \in S$, equivalently, $f \in \operatorname{Ind}_K^G \operatorname{Ran} T$. Thus, (iii) follows as well. $\qquad \square$

1.1.6 Frobenius reciprocity

Theorem 1.1.19 (Frobenius reciprocity) *Let G be a finite group, $K \leq G$ a subgroup, (θ, V) a representation of G and (ρ, W) a representation of K. For $T \in \operatorname{Hom}_G(V, \operatorname{Ind}_K^G W)$ define the linear map $\widehat{T} \colon V \to W$ by setting, for every $v \in V$,*

$$\widehat{T}v = [Tv](1_G). \tag{1.22}$$

Then $\widehat{T} \in \operatorname{Hom}_K(\operatorname{Res}_K^G V, W)$ and the map $T \mapsto \widehat{T}$ is a linear isomorphism between the space of all operators that intertwine (θ, V) with $(\operatorname{Ind}_K^G \rho, \operatorname{Ind}_K^G W)$ and the space of all operators that intertwine the restriction $(\operatorname{Res}_K^G \theta, V)$ of (θ, V) to K with (ρ, W). The corresponding formula is

$$\operatorname{Hom}_G(V, \operatorname{Ind}_K^G W) \cong \operatorname{Hom}_K(\operatorname{Res}_K^G V, W).$$

Proof   We first check that $\widehat{T} \in \operatorname{Hom}_K(\operatorname{Res}_K^G V, W)$; this follows immediately from

$$\begin{aligned}
\widehat{T}\theta(k)v &= [T(\theta(k)v)](1_G) \\
&=_* [\operatorname{Ind}_K^G \rho(k)Tv](1_G) \\
&=_{**} \rho(k)[Tv(1_G)] \\
&= \rho(k)\widehat{T}v
\end{aligned} \tag{1.23}$$

where $k \in K$, $v \in V$; moreover, $=_*$ follows from the fact that T is an intertwiner and $=_{**}$ from the definition of induction.

Conversely, let $S \subseteq G$ be a system of representatives for the left cosets of K in G. Then, given $U \in \operatorname{Hom}_K(\operatorname{Res}_K^G V, W)$, define $\check{U} : V \to \operatorname{Ind}_K^G W$ by setting, for every $v \in V$ and $s \in S$,

$$[\check{U}v](s) = U\theta(s^{-1})v.$$

Again, it is easy to check that $\check{U} \in \operatorname{Hom}_G(V, \operatorname{Ind}_K^G W)$ and, moreover, from

$$[Tv](s) = \left[\operatorname{Ind}_K^G \rho(s^{-1})Tv\right](1_G) = [T\theta(s^{-1})v](1_G) = \widehat{T}\theta(s^{-1})v \tag{1.24}$$

one deduces that $T \mapsto \widehat{T}$ and $U \mapsto \check{U}$ are inverse to one another, thus establishing the required isomorphism. $\square$

In particular we deduce

Corollary 1.1.20   *Let G be a finite group, $K \leq G$ a subgroup and W and V irreducible representations of G and K, respectively. Then the multiplicity of W in $\operatorname{Ind}_K^G V$ equals the multiplicity of V in $\operatorname{Res}_K^G W$.* $\square$

From the point of view of character theory we have

Corollary 1.1.21   *With the same hypotheses as in Theorem 1.1.19,*

$$\frac{1}{|K|}\langle \chi_{\operatorname{Res}_K^G \theta}, \chi_\rho \rangle_K = \frac{1}{|G|}\langle \chi_\theta, \chi_{\operatorname{Ind}_K^G \rho} \rangle_G.$$

$\square$

1.2 Harmonic analysis on a finite homogeneous space

In this section we describe the structure of the permutation representation and of its commutant. Our approach emphasizes the harmonic analytic point of view and is based on [64] and [65].

1.2.1 Frobenius reciprocity for permutation representations

Definition 1.2.1 Let G be a finite group and (σ, V) a representation of G. The *commutant of V* is the algebra $\text{Hom}_G(V, V)$ consisting of all linear operators intertwining V with itself.

Theorem 1.2.2 *Let G be a finite group and (σ, V) a representation of G. Let $V = \bigoplus_{\rho \in I} m_\rho W_\rho$ denote an orthogonal decomposition of V into (not necessarily irreducible) subrepresentations.*

(i) *We have*

$$\text{Hom}_G(V, V) \cong \bigoplus_{\rho, \rho' \in I} m_\rho m_{\rho'} \text{Hom}_G(W_\rho, W_{\rho'}) \qquad (1.25)$$

as vector spaces. In particular,

$$\dim \text{Hom}_G(V, V) = \sum_{\rho, \rho' \in I} m_\rho m_{\rho'} \dim \text{Hom}_G(W_\rho, W_{\rho'}). \qquad (1.26)$$

(ii) *If, in addition, the subrepresentations (ρ, W_ρ), $\rho \in I$, are irreducible and pairwise inequivalent then we have*

$$\text{Hom}_G(V, V) \cong \bigoplus_{\rho \in I} M_{m_\rho, m_\rho}(\mathbb{C}) \qquad (1.27)$$

as algebras. In particular,

$$\dim \text{Hom}_G(V, V) = \sum_{\rho \in I} m_\rho^2. \qquad (1.28)$$

(iii) *Conversely, if (1.28) holds then the subrepresentations (ρ, W_ρ), $\rho \in I$, are irreducible and pairwise inequivalent.*

Proof For every $\rho \in I$ set $V_\rho = m_\rho W_\rho$ and observe that the V_ρ's are G-invariant and pairwise orthogonal (by assumption). Let $T \in \text{Hom}_G(V, V)$. For all $\rho, \rho' \in I$ define a linear map $T_{\rho', \rho} : V_\rho \to V'_\rho$ by setting

$$T_{\rho', \rho} = P_{\rho'} T|_{V_\rho},$$

where $|_{V_\rho}$ denotes the restriction to the subspace V_ρ and $P_{\rho'} : V \to V_{\rho'}$ is the orthogonal projection onto the subspace $V_{\rho'}$. We claim that $T_{\rho',\rho} \in \mathrm{Hom}_G(V_\rho, V_{\rho'})$. Indeed, if $g \in G$ and $v \in V_\rho$, we have

$$
\begin{aligned}
T_{\rho',\rho}\rho(g)v &= P_{\rho'}T|_{V_\rho}\rho(g)v \\
&= P_{\rho'}T\sigma(g)v \\
&= P_{\rho'}\sigma(g)Tv \qquad (\text{since } T \in \mathrm{Hom}_G(V, V)) \\
&= \rho'(g)P_{\rho'}Tv \\
&= \rho'(g)P_{\rho'}T|_{V_\rho}v \\
&= \rho'(g)T_{\rho',\rho}v.
\end{aligned}
$$

It is immediate that the map $T \mapsto (T_{\rho',\rho})_{\rho,\rho'\in I}$ yields a vector space isomorphism

$$
\mathrm{Hom}_G(V, V) \cong \bigoplus_{\rho,\rho'\in I} \mathrm{Hom}_G(V_\rho, V_{\rho'}). \tag{1.29}
$$

Observe now that if $\rho \sim \tau$ and $\rho' \sim \tau'$ then $\mathrm{Hom}_G(W_\rho, W_{\rho'}) \cong \mathrm{Hom}_G(W_\tau, W_{\tau'})$. It then follows from the previous argument that

$$
\mathrm{Hom}_G(V_\rho, V_{\rho'}) = \mathrm{Hom}_G(m_\rho W_\rho, m_{\rho'} W_{\rho'}) \cong m_\rho m_{\rho'}\mathrm{Hom}_G(W_\rho, W_{\rho'}).
$$

From (1.29) we then obtain (1.25) and its immediate consequence (1.26).

Suppose now that the ρ's are irreducible and pairwise inequivalent. By Schur's lemma, we have that $\mathrm{Hom}_G(V_\rho, V_{\rho'})$ is nontrivial if and only if $\rho \sim \rho'$. Then (1.29) becomes the vector space isomorphism

$$
\mathrm{Hom}_G(V, V) \ni T \mapsto (T_{\rho,\rho})_{\rho\in I} \in \bigoplus_{\rho\in I} \mathrm{Hom}_G(V_\rho, V_\rho). \tag{1.30}
$$

Let us show that (1.30) is multiplicative. First observe that, in this case, if $Z \in \mathrm{Hom}_G(V, V)$ and $\rho \in I$ then $Zv = Z_{\rho,\rho}v$ for all $v \in V_\rho$. Thus, if $R, T \in \mathrm{Hom}_G(V, V)$ and $v \in V_\rho$, we have

$$
(RT)_{\rho,\rho}v = RTv = R(T_{\rho,\rho}v) = R_{\rho,\rho}T_{\rho,\rho}v,
$$

showing that $(RT)_{\rho,\rho} = R_{\rho,\rho}T_{\rho,\rho}$. It follows that (1.30) is an algebra isomorphism.

Fix $\rho \in I$; let us show that

$$
\mathrm{Hom}_G(V_\rho, V_\rho) \cong M_{m_\rho,m_\rho}(\mathbb{C}). \tag{1.31}
$$

Let $V_\rho = m_\rho W_\rho = W_\rho^1 \oplus W_\rho^2 \oplus \cdots \oplus W_\rho^{m_\rho}$ denote an orthogonal decomposition of the *isotypic component* V_ρ. Then, using Schur's lemma, we can choose a basis $\{T_{i,j}^\rho : \rho \in I, i, j = 1, 2, \ldots, m_\rho\}$ of $\mathrm{Hom}_G(V_\rho, V_\rho)$ with the following properties:

- $\text{Ker } T_{i,j}^{\rho} = (W_{\rho}^{j})^{\perp}$
- $\text{Ran } T_{i,j}^{\rho} = W_{\rho}^{i}$
- $T_{i,j}^{\rho} T_{j,k}^{\rho} = T_{i,k}^{\rho}.$

Therefore any $T \in \text{Hom}_G(V_{\rho}, V_{\rho})$, can be uniquely written as

$$T = \sum_{i,j=1}^{m_{\rho}} \alpha_{i,j}^{\rho} T_{i,j}^{\rho}$$

with $\alpha_{i,j}^{\rho} \in \mathbb{C}$, and the map

$$T \mapsto (a_{i,j}^{\rho})_{i,j=1}^{m_{\rho}}$$

yields the desired isomorphism (1.31) of algebras. Combining (1.30) and (1.31), we get (1.27) and its immediate consequence (1.28).

Conversely, suppose that (1.28) holds. From (1.26) and the fact that $\dim \text{Hom}_G(V_{\rho}, V_{\rho'}) \geq 1$ if $\rho = \rho'$ (the identity map $\text{Id}_{V_{\rho}} \in \text{Hom}_G(V_{\rho}, V_{\rho})$) and the m_{ρ}'s are nonnegative, we have

$$\sum_{\rho \in I} m_{\rho}^2 = \dim \text{Hom}_G(V, V)$$

$$= \sum_{\rho \in I} m_{\rho}^2 \dim \text{Hom}_G(W_{\rho}, W_{\rho}) + \sum_{\substack{\rho,\rho' \in I \\ \rho \neq \rho' \\ \rho \sim \rho'}} m_{\rho} m_{\rho'} \dim \text{Hom}_G(W_{\rho}, W_{\rho'})$$

$$+ \sum_{\substack{\rho,\rho' \in I \\ \rho \nsim \rho'}} m_{\rho} m_{\rho'} \dim \text{Hom}_G(W_{\rho}, W_{\rho'}),$$

which forces, on the one hand, $\text{Hom}_G(W_{\rho}, W_{\rho'}) = 0$ for all distinct $\rho, \rho' \in I$ such that $\rho \sim \rho'$ (yielding pairwise inequivalence of the ρ's) and, on the other hand, $\dim \text{Hom}_G(W_{\rho}, W_{\rho}) = 1$ (yielding the irreducibility of the ρ's). □

Corollary 1.2.3

(i) *The orthogonal projections onto the isotypic components constitute a basis for the center of* $\text{Hom}_G(V, V)$.

(ii) *An operator T belongs to the center of* $\text{Hom}_G(V, V)$ *if and only if every isotypic component $m_{\rho} W_{\rho}$, $\rho \in I$, constitutes an eigenspace for T.* □

Suppose that G acts transitively on a set X and denote by $(\lambda, L(X))$ the corresponding permutation representation of G. Fix a point $x_0 \in X$ and denote by $K \leq G$ its stabilizer.

Definition 1.2.4   Let G be a group and (σ, V) a G-representation. We denote
by $V^K = \{v \in V : \sigma(k)v = v \text{ for all } k \in K\}$ the subspace of *K-invariant*
vectors in V.

Suppose now that (σ, V) is irreducible, and denote by d_σ its dimension. Also
suppose that V^K is nontrivial. With every $v \in V^K$ we associate a linear map
$T_v : V \to L(X)$ defined by setting

$$(T_v w)(x) = \sqrt{\frac{d_\sigma}{|X|}} \langle w, \sigma(g)v \rangle_V \qquad (1.32)$$

for all $w \in V$ and $x \in X$, where $g \in G$ is such that $gx_0 = x$. Note that
such a group element exists by the transitivity of the action; moreover, (1.32)
is well defined since if $h \in G$ also satisfies $hx_0 = x$ then $h^{-1}gx_0 = x_0$, so that
$h^{-1}g = k \in K$ and therefore $\sigma(g)v = \sigma(hk)v = \sigma(h)\sigma(k)v = \sigma(h)v$. We
have that $T_v \in \mathrm{Hom}_G(V, L(X))$; indeed, for all $g, h \in G$ and $w \in V$,

$$[T_v \sigma(h)w](gx_0) = \sqrt{\frac{d_\sigma}{|X|}} \langle \sigma(h)w, \sigma(g)v \rangle_V$$

$$= \sqrt{\frac{d_\sigma}{|X|}} \langle w, \sigma(h)^{-1}\sigma(g)v \rangle_V$$

$$= \sqrt{\frac{d_\sigma}{|X|}} \langle w, \sigma(h^{-1}g)v \rangle_V$$

$$= [T_v w](h^{-1}gx_0)$$

$$= [\lambda(h)T_v w](gx_0),$$

which shows that $T_v \sigma(h)w = \lambda(h)T_v w$.

Proposition 1.2.5 (Orthogonality relations)   *With the notation above, for all*
$v, u \in V^K$ *and* $w, z \in V$ *one has*

$$\langle T_u w, T_v z \rangle_{L(X)} = \langle w, z \rangle_V \langle v, u \rangle_V. \qquad (1.33)$$

In particular,

(i) *if* $\|v\|_V = 1$ *then* T_v *is an isometric immersion of* V *into* $L(X)$;
(ii) $\mathrm{Im}(T_u) \perp \mathrm{Im}(T_v) \Leftrightarrow u \perp v$.

Proof   Fix $u, v \in V^K$ and define a linear map $R = R_{u,v} : V \to V$ by setting

$$Rw = \frac{1}{|K|} \sum_{g \in G} \langle w, \sigma(g)u \rangle_V v \sigma(g)v \qquad (1.34)$$

for all $w \in V$. It is easy to check that $R \in \text{Hom}_G(V, V)$ and, since V is irreducible, Schur's lemma implies that $R = \alpha I_V$, where $\alpha \in \mathbb{C}$ and $I_V \colon V \to V$ is the identity map on V. Moreover, if $\{w_i : i = 1, 2, \ldots, d_\sigma\}$ is an orthonormal basis for V we have

$$\text{Tr}(R) = \sum_{i=1}^{d_\sigma} \langle Rw_i, w_i \rangle_V$$

$$= \sum_{i=1}^{d_\sigma} \left\langle \frac{1}{|K|} \sum_{g \in G} \langle w_i, \sigma(g)u \rangle_V \sigma(g)v, w_i \right\rangle_V$$

$$= \frac{1}{|K|} \sum_{i=1}^{d_\sigma} \sum_{g \in G} \langle w_i, \sigma(g)u \rangle_V \langle \sigma(g)v, w_i \rangle_V$$

$$= \frac{1}{|K|} \sum_{g \in G} \langle \sigma(g)v, \sigma(g)u \rangle_V$$

$$= \frac{1}{|K|} \sum_{g \in G} \langle v, u \rangle_V$$

$$= |X| \langle v, u \rangle_V.$$

Thus, $\alpha d_\rho = \text{Tr}(\alpha I_V) = \text{Tr}(R) = |X| \langle v, u \rangle_V$, which implies that $R = \frac{|X|}{d_\sigma} \langle v, u \rangle_V I_V$. Therefore, if $w, z \in V$, we have (the overbar denotes the complex conjugate)

$$\langle T_u w, T_v z \rangle_{L(X)} = \frac{1}{|K|} \frac{d_\sigma}{|X|} \sum_{g \in G} \langle w, \rho(g)u \rangle_V \overline{\langle z, \sigma(g)v \rangle_V}$$

$$= \frac{d_\sigma}{|X|} \langle Rw, z \rangle_V$$

$$= \langle w, z \rangle_V \langle v, u \rangle_V.$$

This proves (1.33). Finally, (i) and (ii) follow immediately from (1.33). $\square$

We equip the vector space $\text{Hom}_G(V, L(X))$ with the *normalized Hilbert–Schmidt* scalar product, given by

$$\langle R, T \rangle_{\text{HS}} = \frac{1}{d_\sigma} \sum_{i=1}^{d_\sigma} \langle Rw_i, Tw_i \rangle_V \equiv \frac{1}{d_\sigma} \text{Tr}(R^*S) \tag{1.35}$$

for all $R, T \in \text{Hom}_G(V, L(X))$, where $\{w_i : i = 1, 2, \ldots, d_\sigma\}$ denotes any orthonormal basis in V and $R^* \in \text{Hom}(L(X), V)$ is the adjoint of R.

Theorem 1.2.6 (Frobenius reciprocity for permutation representations)
Let G be a finite group and $K \leq G$ a subgroup and set $X = G/K$. Let (σ, V) be an irreducible G-representation of dimension d_σ and suppose that V^K is nontrivial. For any $v \in V^K$ let $T_v : V \to L(X)$ be as in (1.32). Then the map

$$V^K \ni v \longmapsto T_v \in \operatorname{Hom}_G(V, L(X))$$

is an antilinear isometric vector space isomorphism. In particular, the multiplicity of (σ, V) in $(\lambda, L(X))$ is equal to $\dim V^K$.

Proof   We start by observing that if $\alpha, \beta \in \mathbb{C}$, $u, v \in V^K$, then $T_{\alpha u + \beta v} = \overline{\alpha} T_u + \overline{\beta} T_v$, that is, the map $v \mapsto T_v$ is antilinear. We now show that it is also a bijection. If $T \in \operatorname{Hom}_G(V, L(X))$ then $V \ni w \mapsto (Tw)(x_0) \in \mathbb{C}$ is a linear map and therefore, by the Riesz theorem, there exists a unique $v \in V$ such that $(Tw)(x_0) = \langle w, v \rangle_V$, for all $w \in V$. Therefore, for all $w \in V$ and $g \in G$, we have

$$
\begin{aligned}
[Tw](gx_0) &= [\lambda(g^{-1})Tw](x_0) \\
&= [T\sigma(g^{-1})w](x_0) \qquad \text{(because } T \in \operatorname{Hom}_G(V, L(X))\text{)} \\
&= \langle \sigma(g^{-1})w, v \rangle_V \\
&= \langle w, \sigma(g)v \rangle_V.
\end{aligned} \tag{1.36}
$$

This shows that $T = \sqrt{\frac{|X|}{d_\sigma}} T_v$. Moreover, for all $w \in V$ and $k \in K$ we have from (1.36)

$$\langle w, v \rangle_V = [Tw](x_0) = [Tw](kx_0) = \langle w, \sigma(k)v \rangle_V.$$

It follows that $\sigma(k)v = v$ for all $k \in K$, that is, $v \in V^K$.

Finally, we show that the map is isometric: if $u, v \in V^K$ and $w_1, w_2, \ldots, w_{d_\sigma}$ constitute an orthonormal basis in V then, by (1.33),

$$\langle T_u, T_v \rangle_{\mathrm{HS}} = \frac{1}{d_\sigma} \sum_{j=1}^{d_\sigma} \langle T_u w_j, T_v w_j \rangle_V = \langle v, u \rangle_V.$$

$\square$

Corollary 1.2.7   *The vectors $v_1, v_2, \ldots, v_m$ form an orthogonal basis for V^K if and only if*

$$T_{v_1} V \oplus T_{v_2} V \oplus \cdots \oplus T_{v_m} V$$

is an orthogonal decomposition of the V-isotypic component of $L(X)$. $\square$

Exercise 1.2.8 Use the notation in Theorem 1.1.19 and let $(\iota, \mathbb{C})$ denote the trivial representation of K.

(i) Show that the map $v \mapsto S_v$, $v \in V^K$, defined for all $u \in V$ by

$$S_v(u) = \sqrt{\frac{d_\sigma}{|X|}} \langle u, v \rangle_V,$$

yields a vector space isomorphism from V^K onto $\operatorname{Hom}_K(\operatorname{Res}_K^G V, \mathbb{C})$.

(ii) Show that the composition of the maps $V^K \ni v \mapsto T_v \in \operatorname{Hom}_G(V, L(X))$ (cf. Theorem 1.2.6) and $\operatorname{Hom}_G(V, L(X)) \ni T \mapsto \widehat{T} \in \operatorname{Hom}_K(\operatorname{Res}_K^G V, \mathbb{C})$ (cf. Theorem 1.1.19; here, according to Proposition 1.1.7, we identify $\operatorname{Ind}_K^G \mathbb{C}$ and $L(X)$), namely the map $V^K \ni v \mapsto \widehat{T}_v \in \operatorname{Hom}_K(\operatorname{Res}_K^G V, \mathbb{C})$, coincides with the map $v \mapsto S_v$ in (i).

(iii) Deduce that the following diagram is commutative:

$$V^K \quad \longrightarrow \quad \operatorname{Hom}_G(V, L(X))$$

$$\searrow \qquad\qquad \downarrow$$

$$\operatorname{Hom}_K(\operatorname{Res}_K^G V, \mathbb{C}).$$

1.2.2 Spherical functions

Let G be a finite group. Then the vector space $L(G) = \{f : G \to \mathbb{C}\}$ can be endowed with the structure of an algebra by defining the *convolutional product* $f_1 * f_2$ of two functions $f_1, f_2 \in L(G)$:

$$[f_1 * f_2](g) = \sum_{h \in G} f_1(gh^{-1}) f_2(h)$$

for all $g \in G$.

Now let $K \leq G$ be a subgroup. We denote by $L(G)^K = \{f \in L(G) : f(gk) = f(g) \text{ for all } g \in G, k \in K\}$ and $^K L(G)^K = \{f \in L(G) : f(k_1 g k_2) = f(g) \text{ for all } g \in G, k_1, k_2 \in K\}$ the subsets of $L(G)$ consisting of the *right-K-invariant* and *bi-K-invariant functions* on G, respectively.

Exercise 1.2.9 Show that $L(G)^K$ and $^K L(G)^K$ are indeed subalgebras of $L(G)$.

Our next target is to present a description of the commutant of a permutation representation in terms of bi-K-invariant functions.

Let $X = G/K$ be the homogeneous space consisting of the left cosets of K in G. The group G naturally acts (transitively) on X by left multiplication. We denote by $x_0 \in X$ the coset K. Note that $x_0 \in X$ is fixed by K. Set $L(X) = \{f : X \to \mathbb{C}\}$ and $L(X)^K = \{f \in L(X) : \lambda(k)f = f, \forall k \in K\}$, where λ is the permutation representation.

Exercise 1.2.10   Given $f_1, f_2 \in L(X)$ define $f_1 * f_2 \in L(X)$ by setting

$$[f_1 * f_2](x) = \sum_{ts=g} f_1(tx_0)f_2(sx_0), \tag{1.37}$$

where $x \in X$ and $g \in G$ are such that $gx_0 = x$. Prove that the multiplication in (1.37) is well defined (that is, it does not depend on the particular choice of g) and induces the structure of an algebra on $L(X)$. Show that $L(X)^K$ is a subalgebra of $L(X)$.

With each $f \in L(X)$ we associate $\widetilde{f} \in L(G)$, defined by setting

$$\widetilde{f}(g) = f(gx_0) \tag{1.38}$$

for all $g \in G$.
    Clearly, $\widetilde{f} \in L(G)^K$ for every $f \in L(X)$ and the map $f \mapsto \widetilde{f}$ establishes an algebra isomorphism between $L(X)$ and $L(G)^K$.

Remark 1.2.11   Note that the map $f \to \widetilde{f}$ clearly induces an algebra isomorphism between $L(X)^K$ and ${}^K L(G)^K$.

Theorem 1.2.12   *The commutant $\mathrm{Hom}_G(L(X), L(X))$ is isomorphic to the algebra ${}^K L(G)^K$ of bi-K-invariant functions on G.*

Proof   Let $T \in \mathrm{Hom}_G(L(X), L(X))$ and let $\{\delta_x : x \in X\}$ denote the basis of $L(X)$ consisting of the Dirac functions. Let $(t(x, y))_{x,y \in X}$ be the complex matrix associated with the linear operator T in the given basis, so that $[Tf](x) = \sum_{y \in X} t(x, y)f(y)$ for all $f \in L(X)$ and $x \in X$. Note that, by virtue of the G-invariance of T, we have $t(gx, gy) = t(x, y)$ for all $g \in G$. Consider the function $\psi_T \in L(X)$ defined by $\psi_T(x) = \frac{1}{|K|}t(x, x_0)$ for all $x \in X$. Observe that $\psi_T(kx) = \frac{1}{|K|}t(kx, x_0) = \frac{1}{|K|}t(x, k^{-1}x_0) = \frac{1}{|K|}t(x, x_0) = \psi_T(x)$ for $k \in K, x \in X$, so that $\psi_T \in L(X)^K$. We have, for all $g \in G$ and $f \in L(X)$,

$$\widetilde{Tf}(g) = [Tf](gx_0)$$

$$= \sum_{y \in X} t(gx_0, y) f(y)$$

$$= \frac{1}{|K|} \sum_{h \in G} t(gx_0, hx_0) f(hx_0)$$

$$= \frac{1}{|K|} \sum_{h \in G} t(h^{-1}gx_0, x_0) f(hx_0) \qquad \text{(by the } G\text{-invariance of } T)$$

$$= \sum_{h \in G} \widetilde{\psi}_T(h^{-1}g) \widetilde{f}(h)$$

$$= (\widetilde{f} * \widetilde{\psi}_T)(g).$$

This shows that

$$\widetilde{Tf} = \widetilde{f} * \widetilde{\psi}_T. \tag{1.39}$$

The function $\widetilde{\psi}_T \in {}^K L(G)^K$ is called the *convolutional kernel* of the operator T.

For $\xi \in L(G)$, define $\xi^\sharp \in L(G)$ by setting $\xi^\sharp(g) = \xi(g^{-1})$ for all $g \in G$. Note that the operation $\xi \mapsto \xi^\sharp$ is an involution, that is, $(\xi^\sharp)^\sharp = \xi$, and antimultiplicative, that is,

$$(\xi_1 * \xi_2)^\sharp = \xi_2^\sharp * \xi_1^\sharp \tag{1.40}$$

for all $\xi, \xi_1, \xi_2 \in L(G)$. Finally, note that $\xi^\sharp \in {}^K L(G)^K$ for all $\xi \in {}^K L(G)^K$.

Consider the map $\Psi \colon \operatorname{Hom}_G(L(X), L(X)) \to {}^K L(G)^K$ given by

$$\Psi(T) = (\widetilde{\psi}_T)^\sharp$$

for all $T \in \operatorname{Hom}_G(L(X), L(X))$; let us show that Ψ is the desired algebra isomorphism. Linearity and injectivity are obvious. Moreover, if $\eta \in {}^K L(G)^K$ then the operator $T_\eta \colon L(X) \to L(X)$ defined by

$$\widetilde{T_\eta f} = \widetilde{f} * \eta^\sharp$$

for all $f \in L(X)$ is G-invariant and $\Psi(T_\eta) = \eta$. Indeed, for all $f \in L(X)$ and $g, g_1 \in G$, we have on the one hand

$$[T_\eta \widetilde{\lambda(g)} f](g_1) = [\widetilde{\lambda(g) f} * \eta^\sharp](g_1)$$

$$= \sum_{g_2 \in G} [\widetilde{\lambda(g) f}](g_1 g_2^{-1}) \eta^\sharp(g_2)$$

$$= \sum_{g_2 \in G} \tilde{f}(g^{-1} g_1 g_2^{-1}) \eta^\sharp(g_2)$$

$$= [\tilde{f} * \eta^\sharp](g^{-1} g_1)$$

$$= \widetilde{T_\eta f}(g^{-1} g_1)$$

$$= [\lambda(g)(\widetilde{T_\eta f})](g_1),$$

showing the G-invariance of T_η. On the other hand, for every $f \in L(X)$ we have $\tilde{f} * \eta^\sharp = \widetilde{T_\eta f} = \tilde{f} * \widetilde{\psi_{T_\eta}}$ or, equivalently,

$$\tilde{f} * (\eta^\sharp - \widetilde{\psi_{T_\eta}}) = 0. \tag{1.41}$$

By taking $f = \delta_{gx_0}$ with $g \in G$ in (1.41) one easily gets $\eta^\sharp = \widetilde{\psi_{T_\eta}}$, that is, $\Psi(T_\eta) = \eta$.

To complete the proof we need only to prove that Ψ preserves multiplication. Let $T_1, T_2 \in \operatorname{Hom}_G(L(X), L(X))$. For every $f \in L(X)$ we have, by (1.39) and the definition of the map Ψ,

$$\tilde{f} * \Psi(T_1 \circ T_2)^\sharp = \tilde{f} * \widetilde{\psi_{T_1 \circ T_2}}$$

$$= \widetilde{(T_1 \circ T_2) f}$$

$$= \widetilde{T_1(T_2 f)} = \widetilde{T_2 f} * \Psi(T_1)^\sharp$$

$$= \tilde{f} * \Psi(T_2)^\sharp * \Psi(T_1)^\sharp$$

$$= \tilde{f} * (\Psi(T_1) * \Psi(T_2))^\sharp \qquad \text{(by (1.40))},$$

which shows that $\Psi(T_1 \circ T_2) = \Psi(T_1) * \Psi(T_2)$. $\qquad\square$

Let $I \subseteq \widehat{G}$ denote the set of irreducible representations contained in $L(X)$ (here, and in what follows, we denote by $\widehat{G}$ the *dual* of the group G, that is, a complete set of pairwise inequivalent irreducible representations of the group G). We can summarize Theorem 1.2.12 and the second statement in Theorem 1.2.2 by writing

$$\operatorname{Hom}_G(L(X), L(X)) \cong {}^K L(G)^K \cong \bigoplus_{\rho \in I} M_{m_\rho, m_\rho}(\mathbb{C}), \tag{1.42}$$

where $L(X) = \bigoplus_{\rho \in I} m_\rho V_\rho$ is the decomposition of the permutation representation into its irreducible components (with multiplicities). An irreducible G-representation that appears in the decomposition of the permutation representation is called a *spherical* representation.

Corollary 1.2.13 *With the above notation we have*

$$\sum_{\rho \in I} m_\rho^2 = \text{number of } K\text{-orbits on } X. \tag{1.43}$$

Proof From (1.42) we deduce that

$$\sum_{\rho \in I} m_\rho^2 = \dim \left(\bigoplus_{\rho \in I} M_{m_\rho, m_\rho}(\mathbb{C}) \right) = \dim({}^K L(G)^K).$$

Moreover, we have already seen (cf. Remark 1.2.11) that the algebras ${}^K L(G)^K$ and $L(X)^K$ are isomorphic; in particular $\dim({}^K L(G)^K) = \dim(L(X)^K)$. Finally, we note that a function $f \in L(X)$ belongs to $L(X)^K$ if and only if it is constant on each K-orbit of X. Thus, the dimension of $L(X)^K$ equals the number of such K-orbits and this ends the proof. $\qquad\qquad\square$

In the remaining part of the section, we construct an explicit algebra isomorphism between ${}^K L(G)^K$ and $\bigoplus_{\rho \in I} M_{m_\rho, m_\rho}(\mathbb{C})$. We also introduce two remarkable subalgebras of ${}^K L(G)^K$ that are worthwhile studying, with their relative spherical functions.

Definition 1.2.14 Let (ρ, V_ρ) be a spherical representation. Select an orthonormal basis $B_\rho = \{v_1^\rho, v_2^\rho, \ldots, v_{m_\rho}^\rho\}$ in V_ρ^K, the subspace of K-invariant vectors in V_ρ.

(i) The *spherical matrix coefficients* of (ρ, V_ρ) with respect to B_ρ are the functions $\phi_{i,j}^\rho \in L(G)$, $i, j = 1, 2, \ldots, m_\rho$, defined by

$$\phi_{i,j}^\rho(g) = \langle v_i^\rho, \rho(g) v_j^\rho \rangle_{V_\rho}. \tag{1.44}$$

(ii) The *spherical functions* of (ρ, V_ρ) with respect to B_ρ are the coefficients $\phi_{i,i}^\rho$, $i = 1, 2, \ldots, m_\rho$.

(iii) The *spherical character* of (ρ, V_ρ) is the function $\chi_\rho^K = \sum_{i=1}^{m_\rho} \phi_{i,i}^\rho$.

Exercise 1.2.15 Show that χ_ρ^K is independent of the orthonormal basis B_ρ.

Remark 1.2.16   Note that since the vectors $v_i^\rho, i = 1, 2, \ldots, m_\rho$, are K-invariant then the spherical matrix coefficients $\phi_{i,j}^\rho, i, j = 1, 2, \ldots, m_\rho$, are bi-$K$-invariant. Moreover, by (1.32) we have

$$\phi_{i,j}^\rho(g) = \left[\sqrt{\frac{|X|}{d_\rho}} T_{v_j^\rho} v_i^\rho \right](gx_0)$$

and from Proposition 1.2.5 we deduce that the spherical matrix coefficients $\phi_{i,j}^\rho, i, j = 1, 2, \ldots, m_\rho, \rho \in I$, form an orthogonal basis for $^K L(G)^K$ with $\|\phi_{i,j}^\rho\|_{L(G)}^2 = \frac{|G|}{d_\rho}$ (note that by (1.42) we have $\dim {}^K L(G)^K = \sum_{\rho \in I} m_\rho^2$).

The spherical matrix coefficients satisfy the usual convolutional identity of the matrix coefficients of irreducible representations. We give a proof for the sake of completeness.

Lemma 1.2.17   *With the above notation we have*

$$\phi_{i,j}^\rho * \phi_{h,k}^\sigma = \frac{|G|}{d_\rho} \delta_{j,h} \delta_{\rho,\sigma} \phi_{i,k}^\rho \tag{1.45}$$

for all $i, j = 1, 2, \ldots, m_\rho$ and $h, k = 1, 2, \ldots, m_\sigma$.

Proof   From the orthogonality relations of the spherical matrix coefficients we deduce

$$\phi_{i,j}^\rho * \phi_{h,k}^\sigma(g) = \sum_{t \in G} \phi_{i,j}^\rho(gt) \phi_{h,k}^\sigma(t^{-1})$$

$$= \sum_{t \in G} \langle v_i^\rho, \rho(gt) v_j^\rho \rangle_{V_\rho} \langle v_h^\sigma, \sigma(t^{-1}) v_k^\sigma \rangle_{V_\sigma}$$

$$=_* \sum_{\ell=1}^{d_\rho} \langle v_i^\rho, \rho(g) v_\ell^\rho \rangle_{V_\rho} \sum_{t \in G} \langle v_\ell^\rho, \rho(t) v_j^\rho \rangle_{V_\rho} \overline{\langle v_k^\sigma, \sigma(t) v_h^\sigma \rangle_{V_\sigma}}$$

$$= \sum_{\ell=1}^{d_\rho} \langle v_i^\rho, \rho(g) v_\ell^\rho \rangle_{V_\rho} \delta_{\ell,k} \delta_{j,h} \delta_{\rho,\sigma} \frac{|G|}{d_\rho}$$

$$= \phi_{i,k}^\rho(g) \delta_{j,h} \delta_{\rho,\sigma} \frac{|G|}{d_\rho},$$

where $=_*$ follows from the identity $\rho(g^{-1}) v_i^\rho = \sum_{\ell=1}^{d_\rho} \langle v_i^\rho, \rho(g) v_\ell^\rho \rangle_{V_\rho} v_\ell^\rho$. $\square$

The *spherical Fourier transform* relative to the matrix coefficients (1.44) is the map

$$\begin{array}{ccc} {}^K L(G)^K & \to & \bigoplus_{\rho \in I} M_{m_\rho, m_\rho}(\mathbb{C}) \\ f & \mapsto & \widehat{f} \end{array}$$

where

$$\widehat{f}_{i,j}(\rho) = \langle f, \phi_{i,j}^{\rho} \rangle_{L(G)}.$$

The corresponding inversion formula is given by

$$f(g) = \frac{1}{|G|} \sum_{\rho \in I} d_\rho \sum_{i,j=1}^{m_\rho} \phi_{i,j}^{\rho}(g) \widehat{f}_{i,j}(\rho). \qquad (1.46)$$

Theorem 1.2.18   *The spherical Fourier transform*

$$^{K}L(G)^{K} \to \bigoplus_{\rho \in I} M_{m_\rho, m_\rho}(\mathbb{C})$$

is an isomorphism of algebras.

Proof   Note first that the spherical Fourier transform is a linear isomorphism (whose inverse is given by (1.46)). Thus, it only remains to show the multiplicativity property, namely that

$$\widehat{f * f'} = \widehat{f}\,\widehat{f'} \qquad (1.47)$$

for all $f, f' \in L(K \backslash G / K)$.

By the inversion formula (1.46), given $f, f' \in {}^{K}L(G)^{K}$ we have

$$f * f' = \left(\frac{1}{|G|} \sum_{\rho \in I} d_\rho \sum_{i,j=1}^{m_\rho} \widehat{f}_{i,j}(\rho) \phi_{i,j}^{\rho} \right) * \left(\frac{1}{|G|} \sum_{\sigma \in I} d_\sigma \sum_{h,k=1}^{m_\sigma} \widehat{f'}_{h,k}(\sigma) \phi_{h,k}^{\sigma} \right)$$

$$= \frac{1}{|G|^2} \sum_{\rho,\sigma \in I} d_\rho d_\sigma \sum_{i,j=1}^{m_\rho} \sum_{h,k=1}^{m_\sigma} \widehat{f}_{i,j}(\rho) \widehat{f'}_{h,k}(\sigma) \phi_{i,j}^{\rho} * \phi_{h,k}^{\sigma}$$

$$= \frac{1}{|G|} \sum_{\rho \in I} d_\rho \sum_{i,k=1}^{m_\rho} \sum_{j=1}^{m_\rho} \widehat{f}_{i,j}(\rho) \widehat{f'}_{j,k}(\rho) \phi_{i,k}^{\rho} \qquad \text{(by (1.45))}$$

$$= \frac{1}{|G|} \sum_{\rho \in I} d_\rho \sum_{i,k=1}^{m_\rho} \left(\sum_{j=1}^{m_\rho} \widehat{f}_{i,j}(\rho) \widehat{f'}_{j,k}(\rho) \right) \phi_{i,k}^{\rho}.$$

This shows that $\widehat{f * f'}_{i,k}(\rho) = \sum_{j=1}^{m_\rho} \widehat{f}_{i,j}(\rho) \widehat{f'}_{j,k}(\rho)$, and (1.47) follows. $\square$

Now we examine the effect of a change of basis on the spherical Fourier transform.

Lemma 1.2.19   *Let $U = (u_{ij})_{i,j=1}^{m_\rho}$ be a unitary matrix and let v_i^ρ, $\phi_{i,j}^\rho$ be as above. Then the spherical matrix coefficients with respect to the orthonormal basis*

$$\left\{ \sum_{k=1}^{m_\rho} u_{k,i} v_k^\rho : i = 1, 2, \ldots, m_\rho \right\}$$

are the functions $\xi_{i,j}^\rho$ given by

$$\xi_{i,j}^\rho = \sum_{k,\ell=1}^{m_\rho} u_{i,k} \phi_{k,\ell}^\rho \overline{u_{j,\ell}}.$$

Moreover, if $\widehat{f}(\rho)$ is the spherical Fourier transfrom of a function $f \in {}^K L(G)^K$ with respect to the coefficients $\phi_{i,j}^\rho$ then

$$U^* \widehat{f}(\rho) U$$

is the spherical transform with respect to the coefficients $\xi_{i,j}^\rho$.

Proof   We have

$$\xi_{i,j}^\rho(g) = \sum_{k,\ell=1}^{m_\rho} \langle v_k^\rho u_{k,i}, \rho(g) v_\ell^\rho u_{\ell,j} \rangle v_\rho = \sum_{k,\ell=1}^{m_\rho} u_{k,i} \phi_{k,\ell}^\rho \overline{u_{\ell,j}}$$

and therefore

$$\langle \psi, \xi_{i,j}^\rho \rangle_{L(G)} = \sum_{k,\ell=1}^{m_\rho} \langle \psi, u_{k,i} \phi_{k,\ell}^\rho \overline{u_{\ell,j}} \rangle_{L(G)} = \sum_{k,\ell=1}^{m_\rho} \overline{u_{k,i}} \widehat{\psi_{k,\ell}^\rho} u_{\ell,j}.$$

$\square$

We now consider the ${}^K L(G)^K$-subalgebra

$$\mathcal{A} = \text{span}\{\phi_{i,i}^\rho : \rho \in I, \ i = 1, 2, \ldots, m_\rho\}. \tag{1.48}$$

Clearly, $\mathcal{A}$ depends on the choice of the bases $\{v_1^\rho, v_2^\rho, \ldots, v_{m_\rho}^\rho\}$, $\rho \in I$.

Proposition 1.2.20

(i) $\mathcal{A}$ *is a maximal Abelian subalgebra of ${}^K L(G)^K$.*

(ii) *For $\rho \in I$ and $i = 1, 2, \ldots, m_\rho$, define a linear operator $E_i^\rho : L(X) \to L(X)$ by setting*

$$(E_i^\rho f)(gx_0) = \frac{d_\rho}{|G|} \langle \widetilde{f}, \lambda(g) \phi_{i,i}^\rho \rangle_{L(G)} \equiv \frac{d_\rho}{|G|} [\widetilde{f} * \phi_{i,i}^\rho](g) \tag{1.49}$$

for $f \in L(X)$, $g \in G$ ($\widetilde{f}$ is as in (1.38)). Then E_i^ρ is the orthogonal projection from $L(X)$ onto $T_{v_i^\rho} V_\rho$ ($T_{v_i^\rho} V_\rho$ is as in Proposition 1.2.5).

Proof (i) This follows from the general fact that diagonal matrices form a maximal Abelian subalgebra of a full matrix algebra and that maximal Abelianness is preserved under isomorphisms.

(ii) Extend each basis of V_ρ^K to an orthonormal basis $\{v_1^\rho, v_2^\rho, \ldots, v_{d_\rho}^\rho\}$ of V_ρ, for all $\rho \in I$. Any function $f \in T_{v_k^\sigma} V_\sigma$ is the $T_{v_k^\sigma}$-image of a vector $\sum_{j=1}^{d_\sigma} a_j v_j^\sigma \in V_\sigma^K$, and so it can be expressed as

$$f(hx_0) = \sqrt{\frac{d_\sigma}{|X|}} \sum_{j=1}^{d_\sigma} a_j \langle v_j^\sigma, \sigma(h) v_k^\sigma \rangle_{V_\sigma} = \sqrt{\frac{d_\sigma}{|X|}} \sum_{j=1}^{d_\sigma} a_j \phi_{j,k}^\sigma(h),$$

where $a_j \in \mathbb{C}$ for all $j = 1, 2, \ldots, d_\sigma$ and $h \in G$. Thus,

$$
\begin{aligned}
(E_i^\rho f)(gx_0) &= \frac{d_\rho}{|G|} \sum_{h \in G} f(hx_0) \overline{\phi_{i,i}^\rho(g^{-1}h)} \\
&= \frac{d_\rho}{|G|} \sqrt{\frac{d_\sigma}{|X|}} \sum_{j=1}^{d_\sigma} a_j [\phi_{j,k}^\sigma * \phi_{i,i}^\rho](g) \\
&= \sqrt{\frac{d_\sigma}{|X|}} \sum_{j=1}^{d_\sigma} a_j \delta_{\sigma,\rho} \delta_{k,i} \phi_{j,i}^\sigma(g) \qquad \text{(by (1.45))} \\
&= \begin{cases} f(gx_0) & \text{if } \rho = \sigma \text{ and } i = k \\ 0 & \text{otherwise.} \end{cases}
\end{aligned}
$$

Given a function $\psi \in L(G)$ we define $\psi^\circ \in L(G)$ by setting

$$\psi^\circ(g) = \overline{\psi(g^{-1})}$$

for all $g \in G$.

Lemma 1.2.21

(i) *The spherical Fourier transform of ψ° is*

$$\left(\widehat{\psi^\circ}\right)_{i,j}(\rho) = \overline{\widehat{\psi}_{j,i}(\rho)}$$

for all $\rho \in I$ and $i, j = 1, 2, \ldots, m_\rho$. In matrix terms, $\widehat{\psi^\circ}(\rho)$ is the adjoint of $\widehat{\psi}(\rho)$.

(ii) *If $T \in \mathrm{Hom}_G(L(X), L(X))$ and ψ is its convolution kernel (see (1.39)) then ψ° is the convolution kernel of T^*.*

Proof   (i) Since $\phi_{i,j}^{\rho}(g^{-1}) = \overline{\phi_{j,i}^{\rho}(g)}$, we have

$$
\begin{aligned}
\left(\psi^{\circ}\right)_{i,j}(\rho) &= \langle \psi^{\circ}, \phi_{i,j}^{\rho} \rangle_{L(G)} \\
&= \sum_{g \in G} \overline{\psi(g^{-1})}\, \phi_{i,j}^{\rho}(g) \\
&= \sum_{g \in G} \phi_{j,i}^{\rho}(g)\overline{\psi(g)} \\
&= \overline{\widehat{\psi}_{j,i}(\rho)}.
\end{aligned}
$$

(ii) Recall that we have $\widetilde{Tf} = \tilde{f} * \psi$, for all $f \in L(X)$. Then for $f_1, f_2 \in L(X)$ we get

$$
\begin{aligned}
|K|\langle Tf_1, f_2 \rangle_{L(X)} &\equiv \langle \widetilde{Tf_1}, \tilde{f_2} \rangle_{L(G)} = \sum_{g \in G} \widetilde{f_1} * \psi(g)\, \overline{\tilde{f_2}(g)} \\
&= \sum_{g \in G} \sum_{h \in G} \tilde{f_1}(h)\psi(h^{-1}g)\overline{\tilde{f_2}(g)} \\
&= \sum_{h \in G} \tilde{f_1}(h) \sum_{g \in G} \overline{\tilde{f_2}(g)\psi^{\circ}(g^{-1}h)} \\
&= \langle \tilde{f_1}, \tilde{f_2} * \psi^{\circ} \rangle_{L(G)},
\end{aligned}
$$

and this implies that $\widetilde{T^* f_2} = \tilde{f_2} * \psi^{\circ}$. $\square$

Corollary 1.2.22   *For an operator $T \in \mathrm{Hom}_G(L(X), L(X))$, with convolution kernel ψ, the following conditions are equivalent:*

(i) *T is normal;*
(ii) *ψ and ψ° commute;*
(iii) *T belongs to a maximal commutative subalgebra of the form (1.48).*

Proof   From Lemma 1.2.21 it follows that T and T^* commute if and only if ψ and ψ° commute and this is equivalent to the condition that for each $\rho \in I$ the matrix $\widehat{\psi}(\rho)$ is normal. But $\widehat{\psi}(\rho)$ is normal if and only if it is unitarily diagonalizable. By virtue of Lemma 1.2.19, this is in turn equivalent to the existence of a diagonalizing orthonormal basis $\{v_1^{\rho}, v_2^{\rho}, \ldots, v_{m_\rho}^{\rho}\}$ in V_ρ^K. Invoking the isomorphism (1.42), we end the proof. $\square$

Proposition 1.2.23   *If $T \in \mathrm{Hom}_G(L(X), L(X))$ is normal then we can choose the orthonormal bases $\{v_1^{\rho}, v_2^{\rho}, \ldots, v_{m_\rho}^{\rho}\}$, $\rho \in I$, in such a way that we have an orthogonal decomposition*

$$L(X) = \bigoplus_{\rho \in I} \bigoplus_{i=1}^{m_\rho} T_{v_i^\rho} V_\rho$$

(see Corollary 1.2.7) where each $T_{v_i^\rho} V_\rho$ is an eigenspace of T. Moreover, the eigenvalue corresponding to $T_{v_i^\rho} V_\rho$ is $\widehat{\psi}_{i,i}(\rho)$, where ψ is the convolution kernel of T.

Proof If we choose the bases as at the end of the proof of Corollary 1.2.22, the spherical inversion formula for ψ becomes

$$\psi = \frac{1}{|G|} \sum_{\sigma \in I} d_\sigma \sum_{j=1}^{m_\sigma} \phi_{j,j}^\sigma \widehat{\psi}_{j,j}(\sigma).$$

Moreover, by virtue of (1.49), for all $f \in T_{v_j^\sigma} V_\sigma$ we have

$$\widetilde{f}(g) = \widetilde{E_i^\rho f}(g) = \frac{d_\rho}{|G|} \sum_{h \in G} \widetilde{f}(h) \phi_{i,i}^\rho(h^{-1}g).$$

Therefore, for all $t \in G$, we get

$$\widetilde{Tf}(t) = \widetilde{f} * \psi(t) = \sum_{g \in G} \widetilde{f}(g) \psi(g^{-1}t)$$

$$= \frac{1}{|G|} \sum_{\sigma \in I} d_\sigma \sum_{j=1}^{m_\sigma} \widehat{\psi}_{j,j}(\sigma) \frac{d_\rho}{|G|} \sum_{h \in G} \widetilde{f}(h) \left[\phi_{i,i}^\rho * \phi_{j,j}^\sigma \right](h^{-1}t)$$

$$= \delta_{\rho,\sigma} \delta_{i,j} \widehat{\psi}_{i,i}(\rho) \frac{d_\rho}{|G|} \sum_{h \in G} \widetilde{f}(h) \phi_{i,i}^\rho(h^{-1}t) \qquad \text{(by (1.45))}$$

$$= \widehat{\psi}_{i,i}(\rho) \widetilde{E_i^\rho f}(t)$$

$$= \widehat{\psi}_{i,i}(\rho) \widetilde{f}(t). \qquad \square$$

Consider now the $^K L(G)^K$-subalgebra

$$\mathcal{B} = \text{span}\{\chi_\rho^K : \rho \in I\}. \qquad (1.50)$$

Proposition 1.2.24

(i) $\mathcal{B}$ *is the center of* $^K L(G)^K$.

(ii) *For $\rho \in I$ define a linear operator $E^\rho : L(X) \to L(X)$ by setting*

$$(E^\rho f)(gx_0) = \frac{d_\rho}{|G|} \langle \widetilde{f}, \lambda(g) \chi_\rho^K \rangle_{L(G)} \equiv \frac{d_\rho}{|G|} [\widetilde{f} * \chi_\rho^K](g)$$

for $f \in L(X)$, $g \in G$ ($\widetilde{f}$ is as in (1.38)). Then E^ρ is the orthogonal projection from $L(X)$ onto the isotypic component $m_\rho V_\rho$.

(iii) *If χ_ρ is the character of ρ then the spherical character (see Definition 1.2.14) is given by*

$$\chi_\rho^K(g) = \frac{1}{|K|} \sum_{k \in K} \overline{\chi_\rho(kg)}.$$

Proof Parts (i) and (ii) are consequences of the previous proposition and Corollary 1.2.3. The orthogonal projection of V_ρ onto the subspace V_ρ^K of K-invariant vectors is given by the operator $\frac{1}{|K|} \sum_{k \in K} \rho(k)$; applying the latter to χ_ρ one obtains χ_ρ^K, yielding (iii).   □

Proposition 1.2.25 *An operator $T \in \mathrm{Hom}_G(L(X), L(X))$ belongs to the center of $\mathrm{Hom}_G(L(X), L(X))$ if and only if any isotypic component $m_\rho V_\rho$ constitutes an eigenspace of T. Moreover, if this is the case then the eigenvalue corresponding to $m_\rho V_\rho$ is equal to $\widehat{\psi}_{1,1}(\rho) \equiv \frac{1}{m_\rho} \langle \psi, \chi_\rho^K \rangle$.*

Proof This is just a particular case of Proposition 1.2.23, taking into account that if $T \in \mathrm{Hom}_G(L(X), L(X))$ then also $T^* \in \mathrm{Hom}_G(L(X), L(X))$ and that the center of $M_{m_\rho, m_\rho}(\mathbb{C})$ is constituted by all scalar matrices.   □

Corollary 1.2.26 *If the multiplicity of V_ρ in $L(X)$ is equal to d_ρ then $\chi_\rho^K \equiv \chi_\rho$ and*

$$E^\rho = \frac{d_\rho}{|G|} \sum_{t \in G} \chi_\rho(t) \lambda(t).$$

Proof Suppose that the multiplicity of V_ρ in $L(X)$ is equal to d_ρ. Then $V_\rho = V_\rho^K$ and therefore $\chi_\rho^K = \chi_\rho$. It follows that, for each $f \in L(X)$, we have

$$
\begin{aligned}
E^\rho f(gx_0) &= \frac{d_\rho}{|G|} \langle \widetilde{f}, \lambda(g)\chi_\rho \rangle_{L(G)} && \text{(by Proposition 1.2.24)} \\
&= \frac{d_\rho}{|G|} \sum_{h \in G} f(hx_0) \overline{\chi_\rho(g^{-1}h)} \\
&= \frac{d_\rho}{|G|} \sum_{h \in G} f(hx_0) \overline{\chi_\rho(hg^{-1})} && (\chi_\rho \text{ is central}) \\
&= \frac{d_\rho}{|G|} \sum_{t \in G} f(t^{-1}gx_0) \overline{\chi_\rho(t^{-1})} && (t^{-1} = hg^{-1}) \\
&= \frac{d_\rho}{|G|} \sum_{t \in G} \chi_\rho(t) [\lambda(t)f](gx_0).
\end{aligned}
$$

                                                                                     □

The computational aspects of Corollary 1.2.26 were explored in [21].

1.2.3 The other side of Frobenius reciprocity for permutation representations

In Theorem 1.1.19 Frobenius reciprocity is stated as an explicit isomorphism, namely

$$\text{Hom}_K \left(\text{Res}_K^G V, U \right) \cong \text{Hom}_G \left(V, \text{Ind}_K^G U \right),$$

where $K \leq G$, V is a G-representation and U is a K-representation. The special case in which U is the trivial representation was examined in Theorem 1.2.6. In [7, Corollary 34.1] it is observed that Frobenius reciprocity may be also stated as an explicit isomorphism

$$\text{Hom}_K \left(U, \text{Res}_K^G V \right) \cong \text{Hom}_G \left(\text{Ind}_K^G U, V \right).$$

This formulation of Frobenius reciprocity is particularly useful when the irreducible representations of G are obtained as induced representations; this is the case for a wreath product (see Section 2.4). In this subsection, we examine this equivalent formulation of Frobenius reciprocity in the particular case in which V is a permutation representation.

Let G be again a finite group acting transitively on X and suppose that H is a subgroup of G. Let (ρ, W) be an H-representation. Set $\tau = \text{Ind}_H^G \rho$ and denote by λ the permutation representation of G on X. Let S be a set of representatives for the left cosets of H in G, that is, $G = \bigsqcup_{s \in S} sH$; as usual, we suppose that $1_G \in S$.

Theorem 1.2.27 (Frobenius reciprocity for permutation representations, II) *Let $L \in \text{Hom}_H(W, \text{Res}_H^G L(X))$ and define $\overset{\circ}{L} \colon W^S \to L(X)$ by setting*

$$(\overset{\circ}{L} f)(x) = \frac{1}{\sqrt{|S|}} \sum_{s \in S} [Lf(s)](s^{-1}x), \tag{1.51}$$

for every $f \in W^S = \text{Ind}_H^G W$ and $x \in X$. Then $\overset{\circ}{L} \in \text{Hom}_G(\text{Ind}_H^G W, L(X))$ and the map

$$\text{Hom}_H(W, \text{Res}_H^G L(X)) \longrightarrow \text{Hom}_G(\text{Ind}_H^G W, L(X))$$

$$L \longmapsto \overset{\circ}{L}$$

is an isometric isomorphism.

Proof First observe that if $L \in \text{Hom}_H(W, \text{Res}_H^G L(X))$ then we have

$$[L\rho(h^{-1})w](x) = [\lambda(h^{-1})Lw](x) = [Lw](hx) \tag{1.52}$$

for all $x \in X, h \in H$ and $w \in W$. Let $\tau = \mathrm{Ind}_H^G \rho$, $f \in \mathrm{Ind}_H^G W = W^S$ and $g \in G$. Also, for every $s \in S$ let $t \in S$ and $h \in H$ be the unique elements such that $g^{-1}s = th$. We then have

$$[\overset{\circ}{L} \tau(g)f](x)$$

$$= \frac{1}{\sqrt{|S|}} \sum_{s \in S} \Big(L\{[\tau(g)f](s)\}\Big)(s^{-1}x)$$

$$= \frac{1}{\sqrt{|S|}} \sum_{t \in S} \{L[\rho(h^{-1})f(t)]\}((gth)^{-1}x) \qquad \text{(by Definition 1.1.3)}$$

$$= \frac{1}{\sqrt{|S|}} \sum_{t \in S} [Lf(t)](hh^{-1}t^{-1}g^{-1}x) \qquad \text{(by (1.52))}$$

$$= \frac{1}{\sqrt{|S|}} \sum_{t \in S} [Lf(t)](t^{-1}g^{-1}x)$$

$$= [\overset{\circ}{L} f](g^{-1}x) \qquad \text{(by (1.51))}$$

$$= [\lambda(g) \overset{\circ}{L} f](x).$$

This shows that $\overset{\circ}{L} \in \mathrm{Hom}_G(\mathrm{Ind}_H^G W, L(X))$.

For $w \in W$ define $\delta_w \in W^S$ by setting $\delta_w(s) = \delta_{s,1_G} w$. Then, given $P \in \mathrm{Hom}_G(\mathrm{Ind}_H^G W, L(X))$, we define $\widetilde{P} : W \to L(X)$ by setting

$$\widetilde{P}w = P\delta_w \tag{1.53}$$

for all $w \in W$.

Let us check that $\widetilde{P} \in \mathrm{Hom}_H(W, \mathrm{Res}_H^G L(X))$. Let $h \in H$, $w \in W$ and $s \in S$. We have

$$[\tau(h)\delta_w](s) = \begin{cases} \rho(h)[\delta_w(1_G)] = \rho(h)w & \text{if } s = 1_G \\ \rho(k^{-1})[\delta_w(t)] = 0 & \text{otherwise,} \end{cases}$$

where $t \in S$ and $k \in H$ are the unique elements such that $h^{-1}s = tk$. This shows that

$$\tau(h)\delta_w = \delta_{\rho(h)w}. \tag{1.54}$$

Thus

$$\begin{aligned} \lambda(h)[\widetilde{P}w] &= \lambda(h)[P\delta_w], \\ &= P\tau(h)\delta_w \qquad (P \in \mathrm{Hom}_G(\tau, \lambda)) \\ &= P\delta_{\rho(h)w} \qquad \text{(by (1.54))} \\ &= \widetilde{P}(\rho(h)w) \end{aligned}$$

and therefore $\widetilde{P}$ intertwines W and $\mathrm{Res}_H^G L(X)$. Moreover, for every $L \in \mathrm{Hom}_H(W, \mathrm{Res}_H^G L(X))$, $w \in W$ and $x \in X$, we have (by (1.53) and (1.51)),

$$[\widetilde{\overset{\circ}{L}} w](x) = [\overset{\circ}{L} \, \delta_w](x) = \frac{1}{\sqrt{|S|}} \sum_{s \in S} [L\delta_w(s)](s^{-1}x) = \frac{1}{\sqrt{|S|}}[Lw](x). \quad (1.55)$$

Moreover, first observe that

$$\sum_{s \in S} \tau(s)\delta_{f(s)} = f \qquad (1.56)$$

for all $f \in W^S$. Indeed, for all $t \in S$ we have

$$\left[\sum_{s \in S} \tau(s)\delta_{f(s)}\right](t) = \sum_{s \in S} \rho(h^{-1})\delta_{f(s)}(z) = f(t)$$

where $z \in S$ and $h \in H$ are the unique elements such that $s^{-1}t = zh$ (note that there is only one nonzero summand corresponding to the case $z = 1_G$, which forces $s = t$ and $h = 1_G$). Then, for every $P \in \mathrm{Hom}_G(\mathrm{Ind}_H^G W, L(X))$, $f \in W^S$ and $x \in X$, we have

$$[\overset{\circ}{\widetilde{P}} f](x) = \frac{1}{\sqrt{|S|}} \sum_{s \in S} [\widetilde{P}f(s)](s^{-1}x)$$

$$= \frac{1}{\sqrt{|S|}} \sum_{s \in S} [P\delta_{f(s)}](s^{-1}x)$$

$$= \frac{1}{\sqrt{|S|}} \sum_{s \in S} [\lambda(s)P\delta_{f(s)}](x)$$

$$= \frac{1}{\sqrt{|S|}} \sum_{s \in S} [P\tau(s)\delta_{f(s)}](x)$$

$$= \frac{1}{\sqrt{|S|}} \left(P \sum_{s \in S} [\tau(s)\delta_{f(s)}]\right)(x)$$

$$= \frac{1}{\sqrt{|S|}} P(f)(x) \qquad \text{(by (1.56)).} \qquad (1.57)$$

From (1.55) and (1.57) it follows that the map $P \mapsto \sqrt{|S|}\widetilde{P}$ is the inverse of $L \mapsto \overset{\circ}{L}$ (and, in particular, that $L \mapsto \overset{\circ}{L}$ is an isomorphism).

In order to show that the above map is isometric, let us check that

$$\langle \overset{\circ}{L}_1, \overset{\circ}{L}_2 \rangle_{HS} = \langle L_1, L_2 \rangle_{HS}$$

for all $L_1, L_2 \in \mathrm{Hom}_H(W, \mathrm{Res}_H^G L(X))$. Let $w_1, w_2, \ldots, w_n \in W$ constitute an orthonormal basis of W. Then the functions $\delta_{s,w_i} \in W^S$ defined by

$$\delta_{s,w_i}(t) = \begin{cases} w_j & \text{if } t = s \\ 0 & \text{otherwise} \end{cases}$$

for all $s, t \in S$ and $i = 1, 2, \ldots, n$ constitute an orthonormal basis in W^S (see Remark 1.1.1). Then using (1.51) we deduce that

$$[\overset{\diamond}{L}\,\delta_{s,w_i}](x) = \frac{1}{\sqrt{|S|}}\sum_{t \in S}[L\delta_{s,w_i}(t)](t^{-1}x) = \frac{1}{\sqrt{|S|}}[Lw_i](s^{-1}x) \qquad (1.58)$$

for all $L \in \mathrm{Hom}_H(W, \mathrm{Res}_H^G L(X))$, $s \in S$, $i = 1, 2, \ldots, n$, and $x \in X$. We then have

$$\langle \overset{\diamond}{L_1}, \overset{\diamond}{L_2}\rangle_{\mathrm{HS}} = \sum_{s \in S}\sum_{i=1}^{n}\langle \overset{\diamond}{L_1}\,\delta_{s,w_i}, \overset{\diamond}{L_2}\,\delta_{s,w_i}\rangle_{L(X)}$$

$$= \sum_{s \in S}\sum_{i=1}^{n}\sum_{x \in X}[\overset{\diamond}{L_1}\,\delta_{s,w_i}](x)\overline{[\overset{\diamond}{L_2}\,\delta_{s,w_i}](x)}$$

$$= \frac{1}{|S|}\sum_{s \in S}\sum_{i=1}^{n}\sum_{x \in X}[L_1w_i](s^{-1}x)\overline{[L_2w_i](s^{-1}x)} \qquad \text{(by (1.58))}$$

$$= \sum_{i=1}^{n}\sum_{y \in X}[L_1w_i](y)\overline{[L_2w_i](y)}$$

$$= \sum_{i=1}^{n}\langle L_1w_i, L_2w_i\rangle_{L(X)}$$

$$= \langle L_1, L_2\rangle_{\mathrm{HS}}.$$

$$\square$$

1.2.4 Gelfand pairs

In this subsection we present a special case of the theory illustrated in the previous subsection, namely that of a Gelfand pair.

The theory of Gelfand pairs, originally developed for the infinite case in the setting of Lie groups in the seminal paper by I.M. Gelfand [32], was used, in the finite case, by P. Diaconis [20, 22] to determine the rate of convergence to the stationary distribution of finite Markov chains (we refer to our research expository article [10] and our monograph [11] for a more recent account). Other applications of the theory of finite Gelfand pairs may be found in the monograph by E. Bannai and T. Ito [4] (of a more algebraic combinatorial flavour) and in Ph. Delsarte's thesis [19] (in relation to coding theory). See also A. Terras' monograph [74] as well as the papers [27, 51, 62, 70, 72].

Recall that a representation (σ, V_σ) of a group G is *multiplicity free* if it decomposes into pairwise inequivalent irreducible subrepresentations; in terms of formulas, $V_\sigma = \oplus_{\rho \in I} V_\rho$, where $\rho \not\sim \rho'$ for all distinct $\rho, \rho' \in I$.

Theorem 1.2.28 *Let G be a finite group and $K \leq G$ a subgroup and denote by $X = G/K$ the corresponding homogeneous space. Then the following conditions are equivalent:*

   (i) *the permutation representation $L(X)$ is multiplicity free;*
  (ii) *$\mathrm{Hom}_G(L(X), L(X))$ is commutative;*
 (iii) *$^K L(G)^K$ is commutative;*
  (iv) *for every irreducible G-representation (σ, V), the subspace V^K of the K-invariant vectors is at most one dimensional;*
   (v) *$\mathcal{A} = {}^K L(G)^K$ (where $\mathcal{A}$ is as in (1.48), see also Proposition 1.2.20);*
  (vi) *$\mathcal{B} = {}^K L(G)^K$ (where $\mathcal{B}$ is as in (1.50), see also Proposition 1.2.24).*

Proof Let $L(X) = \oplus_{\rho \in I} m_\rho V_\rho$ be the decomposition of the permutation representation into irreducible components. The equivalence between (i), (ii) and (iii) follows from (1.42). Indeed, the algebra $\oplus_{\rho \in I} M_{m_\rho, m_\rho}(\mathbb{C})$ is Abelian if and only if $m_\rho = 1$ for all $\rho \in I$. Note that, if this is the case, $^K L(G)^K \cong \mathrm{Hom}_G(L(X), L(X)) \cong \mathbb{C}^{|I|}$. The equivalence between (i) and (iv) follows from the Frobenius reciprocity for a permutation representation (Theorem 1.2.6). The equivalence between (iii) and (v) (resp. (vi)) follows from Proposition 1.2.20 (resp. Proposition 1.2.24). $\qquad\square$

If one of the equivalent conditions of the above theorem is satisfied, we say that (G, K) is a *Gelfand pair*. Let G be a finite group acting transitively on a set X. We say that the action gives rise to a Gelfand pair if the permutation representation $(\lambda, L(X))$ is multiplicity free.

Remark 1.2.29 If (G, K) is a Gelfand pair, then for each spherical representation (ρ, V_ρ) there is a unique spherical function ϕ^ρ. In this way the Fourier inversion formula (1.46) becomes

$$f(g) = \frac{1}{|G|} \sum_{\rho \in I} d_\rho \phi^\rho(g) \widehat{f}(\rho), \qquad (1.59)$$

where $\widehat{f}(\rho) = \langle f, \phi^\rho \rangle_{L(G)}$, for all $g \in G$ and $f \in {}^K L(G)^K$.

Example 1.2.30 (Weakly symmetric Gelfand pairs)   Let G be a finite group and $K \leq G$ a subgroup. Suppose that there exists an automorphism τ of G such that

$$g^{-1} \in K\tau(g)K \qquad (1.60)$$

for all $g \in G$. Then (G, K) is a Gelfand pair. To prove this, first observe that if $f \in {}^K L(G)^K$ then by (1.60) we have

$$f(g^{-1}) = f(\tau(g)) \qquad (1.61)$$

for all $g \in G$. Let $f_1, f_2 \in {}^K L(G)^K$. We have

$$
\begin{aligned}
[f_1 * f_2](g) &= \sum_{h \in G} f_1(gh) f_2(h^{-1}) \\
&= \sum_{h \in G} f_1(\tau((gh)^{-1})) f_2(\tau(h)) \qquad \text{(by (1.61))} \\
&= \sum_{t \in G} f_1(\tau(t^{-1})) f_2(\tau(g^{-1}t)) \qquad \text{(by setting } t = gh) \\
&= \sum_{t \in G} f_2(\tau(g^{-1})\tau(t)) f_1(\tau(t^{-1})) \\
&= [f_2 * f_1](\tau(g^{-1})) \\
&= [f_2 * f_1](g) \qquad \text{(by (1.61))}
\end{aligned}
$$

for all $g \in G$. Thus $f_1 * f_2 = f_2 * f_1$, showing that the algebra ${}^K L(G)^K$ is commutative.

We then say that (G, K) is a *weakly symmetric* Gelfand pair.

When the automorphism τ in Example 1.2.30 is equal to the identity, we say that (G, K) is a *symmetric* Gelfand pair. In this case, (1.60) takes the form

$$g^{-1} \in KgK \qquad (1.62)$$

for all $g \in G$.

We now use the following notation. Suppose that a group G acts on a set Y, For two elements $x, y \in Y$ we write $x \sim y$ if there exists $g \in G$ such that $gx = y$ (equivalently, if x and y belong to the same G-orbit, $Gx = Gy$). We also denote by $g(x, y) = (gx, gy)$, $g \in G$ and $x, y \in X$, the *diagonal* action of G on $X \times X$. Finally, we say that the orbits of G on $X \times X$ (with respect to the diagonal action) are *symmetric* provided that for all $x, y \in X$ one has $(x, y) \sim (y, x)$.

Proposition 1.2.31 *Let X be a finite set and G a group acting transitively on it. Let $x_0 \in X$ and $K = \text{Stab}_G(x_0) = \{k \in G : kx_0 = x_0\}$ be a point of X and its stabilizer, respectively. Then the following conditions are equivalent:*

(i) *the orbits of G on $X \times X$ are symmetric;*
(ii) *(G, K) is a symmetric Gelfand pair.*

Proof (i) $\Rightarrow$ (ii). Suppose that the orbits of G on $X \times X$ are symmetric. Let $g \in G$. As $(x_0, g^{-1}x_0) = g^{-1}(gx_0, x_0) \sim (gx_0, x_0) \sim_{(i)} (x_0, gx_0)$, there exists $k \in G$ such that $k(x_0, g^{-1}x_0) = (x_0, gx_0)$. This is equivalent to $kx_0 = x_0$ (so that $k \in K$) and $kg^{-1}x_0 = gx_0$. The last condition then gives $g^{-1}kg^{-1}x_0 = x_0$, that is, $g^{-1}kg^{-1} \in K$. We then have (1.62) so that (G, K) is a symmetric Gelfand pair.

(ii) $\Rightarrow$ (i). Suppose that (G, K) is a symmetric Gelfand pair. Let $x, y \in X$. Then, by transitivity of the action, we can find $t, s \in G$ such that $x = tx_0$ and $y = tsx_0$. Moreover, by (1.62), we can find $k_1, k_2 \in K$ such that $s^{-1} = k_1 s k_2$. We then have

$$(x, y) = t(x_0, t^{-1}y) \sim (x_0, t^{-1}y) = (x_0, sx_0)$$

$$= s(s^{-1}x_0, x_0) \sim (s^{-1}x_0, x_0) = (k_1 s k_2 x_0, x_0)$$

$$= (k_1 s x_0, x_0) = k_1(sx_0, k_1^{-1}x_0) = k_1(sx_0, x_0)$$

$$\sim (sx_0, x_0) = (t^{-1}y, x_0) \sim t(t^{-1}y, x_0) = (y, x).$$

This shows that the orbits of G on $X \times X$ are symmetric.   □

Example 1.2.32 (2-point homogeneous Gelfand pairs) Let G be a finite group acting isometrically on a metric space (X, d). We say that the action is *2-point homogeneous* (or *distance-transitive*) if, for all $(x_1, y_1), (x_2, y_2)$ in $X \times X$ such that $d(x_1, y_1) = d(x_2, y_2)$, there exists $g \in G$ such that $gx_1 = x_2$ and $gy_1 = y_2$; equivalently,

$$d(x_1, y_1) = d(x_2, y_2) \implies (x_1, y_1) \sim (x_2, y_2).$$

Note that, in particular, a 2-point homogeneous action is transitive. Since the distance function d is symmetric, that is, $d(x, y) = d(y, x)$ for all $x, y \in X$, we deduce that the G-orbits on $X \times X$ are symmetric. Thus, by Proposition 1.2.31, every 2-point homogeneous action gives rise to a symmetric Gelfand pair.

Definition 1.2.33 Suppose that G acts on X. The action is *doubly transitive* if for all $(x_1, x_2), (y_1, y_2) \in (X \times X) \setminus \{(x, x) : x \in X\}$ there exists $g \in G$ such that $gx_i = y_i$ for $i = 1, 2$.

Exercise 1.2.34   Suppose that G acts transitively on X. Set $W_0 = \{f : X \to \mathbb{C}, \text{constant}\}$ and $W_1 = \{f : X \to \mathbb{C}, \sum_{x \in X} f(x) = 0\}$. Prove that $L(X) = W_0 \oplus W_1$ is the decomposition of the permutation representation into irreducibles if and only if G acts doubly transitively on X.

In the context of Gelfand pairs, we can reformulate Corollary 1.2.13 as follows.

Corollary 1.2.35   *Let (G, K) be a Gelfand pair and let $I \subseteq \widehat{G}$ denote the set of irreducible representations contained in the corresponding permutation representation $L(X)$. Then*

$$|I| = \text{number of } K\text{-orbits on } X. \tag{1.63}$$

$\square$

We end this section with the following useful criterion for Gelfand pairs, which is, in some sense, a converse to Corollary 1.2.35.

Theorem 1.2.36   *Let G be a finite group and $K \leq G$ a subgroup and denote by $X = G/K$ the corresponding homogeneous space. Suppose we have a decomposition*

$$L(X) = \oplus_{t \in T} Z_t$$

of the permutation representation into G-subrepresentations with $|T|$ equal to the number of K-orbits on X. Then the Z_t's are irreducible and (G, K) is a Gelfand pair.

Proof   Refine if necessary the decomposition $L(X)$ of the Z_t's into irreducibles to obtain $L(X) = \sum_{\rho \in I} m_\rho V_\rho$ (where the V_ρ's are pairwise nonequivalent). Then $|T| \leq |I| \leq \sum_{\rho \in I} m_\rho^2$ and Corollary 1.2.13 forces $|T| = |I|$ and $m_\rho = 1$ for all $\rho \in I$, concluding the proof. $\square$

1.3   Clifford theory

In this section we analyze the relation between representations of a given group and those of its normal subgroups. These results were obtained by Alfred H. Clifford in 1937 [16]. We follow Huppert's monograph [35] quite closely but our approach, based on our research-expository paper [13], rests on an explicit analysis of the representation spaces and their decompositions rather than on calculations with characters (as in [35]). This functional framework is more suitable for applications in harmonic analysis problems. We also avoid a direct application of Mackey's lemma (Theorem 4.1 in [12]) but carefully examine the representations involved. Other excellent expositions on Clifford theory are the monographs by Isaacs [36] and by Curtis and Reiner [17, 18]. The latter also uses an explicit analysis of the representation spaces but with a more algebraic language. See also the monographs by Grove [34], Berkovich and Zhmud [5, 6] and Dornhoff [24].

As an application of Clifford theory, we prove a general form of the so-called *little group method*, which provides an useful way to get a complete list of irreducible representations for a wide class of groups. In particular, in the next chapter we shall use it to obtain the representation theory of wreath products.

1.3.1 Clifford correspondence

In this section we introduce the main definitions and present Clifford correspondence (Theorem 1.3.6).

Let G be a finite group and $N \trianglelefteq G$ a normal subgroup of G. Denote by $\widehat{G}$ (resp. $\widehat{N}$) a set of pairwise inequivalent irreducible representations of G (resp. N), which, by an abuse of notation, we also identify with the set of all equivalence classes of irreducible representations of G (resp. N).

In the following, we use the following notation. Given two representations σ and ρ we write $\sigma \preceq \rho$, and we say that σ is *contained* in ρ, if σ is a subrepresentation of ρ.

Definition 1.3.1 *Let $\sigma \in \widehat{N}$ and $g \in G$.*

(i) *We set*

$$\widehat{G}(\sigma) = \{\theta \in \widehat{G} : \sigma \preceq \mathrm{Res}_N^G(\theta)\} \subset \widehat{G}.$$

(ii) *The g-conjugate of σ is the representation ${}^g\sigma \in \widehat{N}$ defined by*

$$
{}^g\sigma(n) = \sigma(g^{-1}ng) \tag{1.64}
$$

for all $n \in N$.

(iii) *The subgroup*

$$I_G(\sigma) = \{g \in G : {}^g\sigma \sim \sigma\} \leq G$$

is called the inertia group of $\sigma \in \widehat{N}$.

It is easy to see that (1.64) defines a left action of G on $\widehat{N}$, that is, ${}^{g_1g_2}\sigma = {}^{g_1}({}^{g_2}\sigma)$ for all $g_1, g_2 \in G$ and $\sigma \in \widehat{N}$. Moreover, if $\sigma_1 \sim \sigma_2$ and $g \in G$ then ${}^g\sigma_1 \sim {}^g\sigma_2$, so that the action preserves the equivalence relation. Thus $I_G(\sigma)$ is the stabilizer of the equivalence class $[\sigma]$ of $\sigma \in \widehat{N}$ in G under this action. In particular, we have

$$I_G({}^g\sigma) = g^{-1}I_G(\sigma)g \tag{1.65}$$

for all $g \in G$ and $\sigma \in \widehat{N}$. Observe that $I_G(\sigma)$ contains the subgroup N. Indeed, if $n, n_1 \in N$ we have $^{n_1}\sigma(n) = \sigma(n_1)^{-1}\sigma(n)\sigma(n_1)$ and therefore $^{n_1}\sigma \sim \sigma$. We then denote by Q a set of representatives for the left N-cosets in $I_G(\sigma)$ such that that $1_G \in Q$. We thus have

$$I_G(\sigma) = \coprod_{q \in Q} qN. \tag{1.66}$$

Similarly, we denote by R a set of representatives for the left $I_G(\sigma)$-cosets in G such that that $1_G \in R$, so that

$$G = \coprod_{r \in R} r I_G(\sigma). \tag{1.67}$$

Then $\{[^g\sigma] : g \in G\} \equiv \{[^r\sigma] : r \in R\}$ and the representations $^r\sigma, r \in R$, are pairwise nonequivalent.

Moreover, $T = RQ$ is a set of representatives for the left N-cosets in G. From (1.66) and (1.67) we deduce that

$$G = \coprod_{r \in R} r I_G(\sigma) = \coprod_{r \in R} \coprod_{q \in Q} rqN = \coprod_{t \in T} tN.$$

Note that $1_G \in T$ and that $R \subseteq T$ (resp. $Q \subseteq T$), since $1_G \in R$ (resp. $1_G \in Q$).

Theorem 1.3.2 *Suppose that $N \trianglelefteq G$ and let $\sigma \in \widehat{N}$ and $\theta \in \widehat{G}(\sigma)$. If R, Q and T are as above then, setting $d = [I_G(\sigma) : N] = |Q|$ and denoting by ℓ the multiplicity of σ in $\mathrm{Res}_N^G \theta$, we have that:*

(i) $\mathrm{Res}_N^G \mathrm{Ind}_N^G \sigma = \bigoplus_{t \in T} {}^t\sigma = d \bigoplus_{r \in R} {}^r\sigma$ *expresses the decomposition of* $\mathrm{Res}_N^G \mathrm{Ind}_N^G \sigma$ *into irreducible subrepresentations;*

(ii) $\mathrm{Hom}_G(\mathrm{Ind}_N^G \sigma, \mathrm{Ind}_N^G \sigma) \cong \mathbb{C}^d$;

(iii) $\mathrm{Res}_N^G \theta \cong \ell \bigoplus_{r \in R} {}^r\sigma.$

Proof (i) Let V_σ be the representation space of σ and let $t \in T$. Set $Z_t = \{f \in V_\sigma^T : f(s) = 0 \text{ for all } s \in T \text{ such that } s \neq t\}$. Then the linear map $L_t : V_\sigma \to Z_t$ defined by $[L_t v](s) = \delta_{t,s} v$, for all $s \in S$, is bijective. Moreover, Z_t is $(\mathrm{Res}_N^G \mathrm{Ind}_N^G \sigma)$-invariant and intertwines the N-representations $(\mathrm{Res}_N^G \mathrm{Ind}_N^G \sigma)|_{Z_t}$ and $^t\sigma$. In order to verify these facts, let $n \in N$, $v \in V_\sigma$, $f \in Z_t$ and $s \in T$. Then we can write $n^{-1}s = s(s^{-1}n^{-1}s) = sm$, where $m = s^{-1}n^{-1}s \in N$ as N is normal in G. We now have

$$[(\text{Res}_N^G \text{Ind}_N^G \sigma)(n)f](s) = [\text{Ind}_N^G \sigma(n)f](s)$$
$$= \sigma(m^{-1})f(s)$$
$$= \delta_{s,t}\sigma(m^{-1})f(t)$$
$$= \delta_{s,t}[(\text{Res}_N^G \text{Ind}_N^G \sigma)(n)f](s),$$

that is, $(\text{Res}_N^G \text{Ind}_N^G \sigma)(n)f \in Z_t$ for all $f \in Z_t$ and $n \in N$. Similarly,

$$[(\text{Res}_N^G \text{Ind}_N^G \sigma)(n)L_t v](s) = [(\text{Ind}_N^G \sigma)(n)L_t v](s)$$
$$= \sigma(m^{-1})[(L_t v)(s)]$$
$$= \delta_{t,s}\sigma(m^{-1})v$$
$$= \delta_{t,s}\sigma(s^{-1}ns)v$$
$$= \delta_{t,s}\sigma(t^{-1}nt)v$$
$$= \delta_{t,s}{}^t\sigma(n)v$$
$$= L_t[{}^t\sigma(n)v](s).$$

This shows that $(\text{Res}_N^G \text{Ind}_N^G \sigma)(n)L_t = L_t{}^t\sigma(n)$, that is,

$$L_t \in \text{Hom}_N((\text{Res}_N^G \text{Ind}_N^G \sigma)|_{Z_t}, Z_t), ({}^t\sigma, V_\sigma))$$

(observe that $V_\sigma = V_{{}^t\sigma}$). As $\text{Res}_N^G \text{Ind}_N^G V_\sigma = \bigoplus_{t \in T} Z_t$, we deduce that the operator $T = \bigoplus_{t \in T} L_t$ is an isomorphism between $\text{Res}_N^G \text{Ind}_N^G \sigma$ and $\bigoplus_{t \in T} {}^t\sigma$. We finally have

$$\bigoplus_{t \in T} {}^t\sigma = \bigoplus_{r \in R} \bigoplus_{q \in Q} {}^{rq}\sigma = |Q| \bigoplus_{r \in R} {}^r\sigma.$$

(ii) From the previous result we deduce that the multiplicity of σ in $\text{Res}_N^G \text{Ind}_N^G \sigma$ is equal to d. Therefore, by Frobenius reciprocity (Theorem 1.1.19) we have

$$\text{Hom}_G(\text{Ind}_N^G \sigma, \text{Ind}_N^G \sigma) = \text{Hom}_N(\sigma, \text{Res}_N^G \text{Ind}_N^G \sigma) \cong \mathbb{C}^d.$$

(iii) Let $g \in G$ and $\Phi \in \text{Hom}_N(\sigma, \text{Res}_N^G \theta)$. Then $\theta(g)\Phi \in \text{Hom}_N({}^g\sigma, \text{Res}_N^G \theta)$. Indeed, for $n \in N$ we have

$$\theta(g)\Phi{}^g\sigma(n) = \theta(g)\Phi\sigma(g^{-1}ng)$$
$$= \theta(g)\theta(g^{-1}ng)\Phi$$
$$= \theta(n)\theta(g)\Phi.$$

Moreover, the map

$$\begin{array}{ccc} \text{Hom}_N(\sigma, \text{Res}_N^G \theta) & \longrightarrow & \text{Hom}_N({}^g\sigma, \text{Res}_N^G \theta) \\ \Phi & \longmapsto & \theta(g)\Phi \end{array} \qquad (1.68)$$

is a linear isomorphism. This follows immediately after observing that $g^{-1}(g\sigma) = \sigma$, so that the map

$$\mathrm{Hom}_N(g\sigma, \mathrm{Res}_N^G\theta) \quad\longrightarrow\quad \mathrm{Hom}_N(\sigma, \mathrm{Res}_N^G\theta)$$
$$\Psi \qquad\qquad\longmapsto\qquad \theta(g^{-1})\Psi$$

is the inverse of (1.68). Therefore every $g\sigma$ has multiplicity ℓ in $\mathrm{Res}_N^G\theta$, that is, $\mathrm{Res}_N^G\theta \succeq \ell\bigoplus_{r\in R} {}^r\sigma$. By Frobenius reciprocity, $\mathrm{Ind}_N^G\sigma$ contains exactly ℓ copies of θ so that every irreducible subrepresentation in $\mathrm{Res}_N^G\theta$ is also a subrepresentation of $\mathrm{Res}_N^G\mathrm{Ind}_N^G\sigma$. But, by (i), the latter contains only irreducible subrepresentations of the form ${}^r\sigma$, and this ends the proof.   □

Corollary 1.3.3   *With the notation of Theorem 1.3.2, we have that*

(i) $\mathrm{Ind}_N^G\sigma$ *is irreducible if and only if* $I_G(\sigma) = N$;
(ii) *if* $\sigma, \sigma' \in \widehat{N}$ *and* $I_G(\sigma) = N = I_G(\sigma')$ *then* $\mathrm{Ind}_N^G\sigma \sim \mathrm{Ind}_N^G\sigma'$ *if and only if* σ *and* σ' *are conjugate (that is, there exists* $g \in G$ *such that* $\sigma' = {}^g\sigma$).

Proof   The first statement follows from (ii) in Theorem 1.3.2 combined with Schur's lemma.

Suppose now that $I_G(\sigma) = N = I_G(\sigma')$. By (i) in Theorem 1.3.2 we have $\mathrm{Res}_N^G\mathrm{Ind}_N^G\sigma = \bigoplus_{r\in R} {}^r\sigma$. Then Frobenius reciprocity implies that

$$\mathrm{Hom}_N(\sigma', \bigoplus_{r\in R} {}^r\sigma) \cong \mathrm{Hom}_N(\sigma', \mathrm{Res}_N^G\mathrm{Ind}_N^G\sigma) \cong \mathrm{Hom}_G(\mathrm{Ind}_N^G\sigma', \mathrm{Ind}_N^G\sigma).$$

Since $\mathrm{Ind}_N^G\sigma'$ and $\mathrm{Ind}_N^G\sigma$ are G-irreducible and σ' and ${}^r\sigma$, $r \in R$, are N-irreducible, by Schur's lemma we deduce that $\mathrm{Ind}_N^G\sigma' \sim \mathrm{Ind}_N^G\sigma$ if and only if there exists $r \in R$ such that $\sigma' \sim {}^r\sigma$.   □

Definition 1.3.4   For $\sigma \in \widehat{N}$ and $\theta \in \widehat{G}(\sigma)$, the number

$$\ell = \dim \mathrm{Hom}_N(\sigma, \mathrm{Res}_N^G\theta) \qquad\qquad (1.69)$$

is called the *inertia index* of θ with respect to N.

Note that, a priori, given $\sigma \in \widehat{N}$ and $\theta \in \widehat{G}(\sigma)$ the number (1.69) is also defined in terms of the representation σ rather than only the subgroup N. However, Theorem 1.3.2(iii) guarantees that ℓ is, in fact, independent of σ.

Lemma 1.3.5   *Let* $N \trianglelefteq G$ *be a normal subgroup of* G *and* $\sigma \in \widehat{N}$. *Denote by* $I = I_G(\sigma)$ *the inertia group of* σ. *Let*

$$\mathrm{Ind}_N^I\sigma = \bigoplus_{\psi\in\widehat{I}(\sigma)} m_\psi \psi$$

be the decomposition of $\mathrm{Ind}_N^I\sigma$ *into* I-*irreducible representations (*$m_\psi > 0$ *is the multiplicity of* ψ). *Then the following hold.*

(i) $\mathrm{Ind}_N^G \sigma = \bigoplus_{\psi \in \widehat{I}(\sigma)} m_\psi \, \mathrm{Ind}_I^G \psi$ *is the decomposition of* $\mathrm{Ind}_N^G \sigma$ *into its G-irreducible components (that is, the* $\mathrm{Ind}_I^G \psi$*'s are G-irreducible and pairwise inequivalent).*

(ii) *If* $\theta \in \widehat{G}(\sigma)$ *then* $\theta = \mathrm{Ind}_I^G \psi$ *for some (unique)* $\psi \in \widehat{I}(\sigma)$.

Proof First note that, from the definition above and Frobenius reciprocity, we have $\widehat{I}(\sigma) = \{\psi \in \widehat{I} : \psi \preceq \mathrm{Ind}_N^I \sigma\}$. (i) By the transitivity (Proposition 1.1.10) and additivity (Proposition 1.1.11) of induction,

$$\mathrm{Ind}_N^G \sigma = \mathrm{Ind}_I^G \mathrm{Ind}_N^I \sigma = \bigoplus_{\psi \in \widehat{I}(\sigma)} m_\psi \, \mathrm{Ind}_I^G \psi.$$

Moreover,

$$\mathrm{Hom}_G(\mathrm{Ind}_N^G \sigma, \mathrm{Ind}_N^G \sigma) = \mathbb{C}^d = \mathrm{Hom}_I(\mathrm{Ind}_N^I \sigma, \mathrm{Ind}_N^I \sigma),$$

where the first equality is exactly Theorem 1.3.2 (ii) and the second follows from the same result with G replaced by I (note that, indeed, I coincides with the inertia group of σ in I). Then, by the commutant theorem (Theorem 1.2.2), we have

$$\sum_{\psi \in \widehat{I}(\sigma)} m_\psi^2 = d = \sum_{\psi, \eta \in \widehat{I}(\sigma)} m_\psi m_\eta \, \dim \mathrm{Hom}_G(\mathrm{Ind}_I^G \psi, \mathrm{Ind}_I^G \eta).$$

Therefore $\dim \mathrm{Hom}_G(\mathrm{Ind}_I^G \psi, \mathrm{Ind}_I^G \eta) = \delta_{\psi, \eta}$ and (i) follows.

(ii) This is an immediate consequence of Frobenius reciprocity: if $\sigma \preceq \mathrm{Res}_N^G \theta$ then $\theta \preceq \mathrm{Ind}_N^G \sigma$ and therefore $\theta = \mathrm{Ind}_I^G \psi$ for some $\psi \in \widehat{I}(\sigma)$. $\square$

Theorem 1.3.6 (Clifford correspondence) *Let* $N \trianglelefteq G$ *and* $\sigma \in \widehat{N}$, *set* $I = I_G(\sigma)$ *and let* $\widehat{I}(\sigma)$ *be as before. Then:*

(i) *the map*

$$\begin{array}{ccc} \widehat{I}(\sigma) & \longrightarrow & \widehat{G}(\sigma) \\ \psi & \longmapsto & \mathrm{Ind}_I^G \psi \end{array} \tag{1.70}$$

is a bijection;

(ii) *the inertia index of* $\psi \in \widehat{I}(\sigma)$ *with respect to N coincides with the inertia index of* $\mathrm{Ind}_I^G \psi$ *with respect to N, and they are equal to* m_ψ *(the multiplicity of* ψ *in* $\mathrm{Ind}_N^I \sigma$*);*

(iii) $\mathrm{Res}_N^I \psi = m_\psi \sigma$.

Proof Part (i) is just a reformulation of Lemma 1.3.5. Part (ii) follows from Frobenius reciprocity: if $\psi \in \widehat{I}(\sigma)$ then the inertia index of ψ with respect to N is equal to m_ψ and coincides with the inertia index of $\mathrm{Ind}_I^G \psi$ with respect to N (by Lemma 1.3.5(i)). Finally, (iii) is given by an application of Theorem 1.3.2(iii), with G replaced by I:

$$\mathrm{Res}^I_N \psi = \underbrace{\sigma \oplus \sigma \oplus \cdots \oplus \sigma}_{m_\psi \text{ times}}.$$

$\square$

Remark 1.3.7 Note that if $I_G(\sigma) = G$ (that is, $^g\sigma \sim \sigma$ for all $g \in G$) then the first two statements of the Clifford correspondence are trivial. Indeed, if this is the case, we have $\widehat{I}(\sigma) = \widehat{G}(\sigma)$ and the correspondence (1.70) becomes the identity map $\psi \mapsto \mathrm{Ind}^G_I \psi = \psi$. Moreover, the inertia index of ψ with respect to N, which by definition equals the dimension of $\mathrm{Hom}_N(\sigma, \mathrm{Res}^G_N \psi)$, is in turn equal by Frobenius reciprocity to m_ψ, the dimension of $\mathrm{Hom}_G(\mathrm{Ind}^G_N \sigma, \psi)$. Finally, note that if $\theta \in \widehat{G}$ we have $\sigma := \mathrm{Res}^G_N \theta \in \widehat{N}$ if and only if $I_G(\sigma) = G$ and the inertia index of θ with respect to N is equal to 1 (see Theorem 1.3.2(iii)).

We summarize the results obtained in Table 1.1.
Now let $\psi \in \widehat{G/N}$. We define the *inflation* $\overline{\psi} \in \widehat{G}$ of ψ by setting

$$\overline{\psi}(g) = \psi(gN) \qquad \forall g \in G. \tag{1.71}$$

Note that $\overline{\psi}$ is indeed a representation (of G) since it equals the composition of the quotient homomorphism $G \to G/N$ and the representation $\psi \in \widehat{G/N}$.

Table 1.1

Induction
$\mathrm{Ind}^G_N \sigma = \bigoplus_{\psi \in \widehat{I}(\sigma)} m_\psi \mathrm{Ind}^G_I \psi$
$\uparrow$
$\mathrm{Ind}^I_N \sigma = \bigoplus_{\psi \in \widehat{I}(\sigma)} m_\psi \psi$
$\uparrow$
$\sigma \in \widehat{N}$

Restriction from I
$\psi \in \widehat{I}(\sigma)$
$\downarrow$
$\mathrm{Res}^I_N \psi = m_\psi \sigma$

Restriction from G
$\mathrm{Ind}^G_I \psi \in \widehat{G}(\sigma)$
$\downarrow$
$\mathrm{Res}^G_N \mathrm{Ind}^G_I \psi = m_\psi \bigoplus_{r \in R} {}^r\sigma$

Moreover, if ψ is irreducible then $\overline{\psi}$ is irreducible as well and if $\psi = \psi_1 \oplus \psi_2$ then $\overline{\psi} = \overline{\psi_1} \oplus \overline{\psi_2}$.

Recall that the *left-regular representation* of a group G is the G-representation $(\lambda, L(G))$, where $[\lambda(g)f](h) = f(g^{-1}h)$ for all $g, h \in G$ and $f \in L(G)$ (this is a particular case of the permutation representation (see Definition 1.1.6) corresponding to when G acts on itself by left multiplication).

Example 1.3.8 Suppose that ψ is the left-regular representation of the group G/N. Then $\overline{\psi}$ is exactly the permutation representation of G over $X = G/N$. In other words (cf. Proposition 1.1.7),

$$\overline{\psi} \sim \operatorname{Ind}_N^G \iota_N \qquad (1.72)$$

where ι_N is the trivial representation of N.

Theorem 1.3.9 (Gallagher [30]) *Let $N \trianglelefteq G$ and $\theta \in \widehat{G}$ and suppose that $\sigma := \operatorname{Res}_N^G \theta \in \widehat{N}$ (see Remark 1.3.7). For $\psi \in \widehat{G/N}$, denote by d_ψ the dimension of ψ. Then the following hold:*

(i)

$$\operatorname{Ind}_N^G \sigma = \bigoplus_{\psi \in \widehat{G/N}} d_\psi (\theta \otimes \overline{\psi}) \qquad (1.73)$$

  where the $\theta \otimes \overline{\psi}$ are irreducible and pairwise nonequivalent;

(ii) *the inertia index of $\theta \otimes \overline{\psi}$ with respect to N is equal to d_ψ;*

(iii) *if $\tau \in \widehat{G}$ and $\operatorname{Res}_N^G \tau = \sigma$ then $\tau \sim \theta \otimes \overline{\psi}$ for some $\psi \in \widehat{G/N}$ with $d_\psi = 1$.*

Proof By Proposition 1.1.15 we have

$$\operatorname{Ind}_N^G \sigma = \operatorname{Ind}_N^G (\sigma \otimes \iota_N) = \operatorname{Ind}_N^G [(\operatorname{Res}_N^G \theta) \otimes \iota_N] = \theta \otimes \operatorname{Ind}_N^G \iota_N = \theta \otimes \overline{\lambda},$$

where ι_N denotes the trivial representation of N and $\overline{\lambda}$ is the inflation of the regular representation λ of G/N (see (1.72)). Recalling (see for instance [11, Theorem 3.7.11(iii)]) that $\lambda = \oplus_{\psi \in \widehat{G/N}} d_\psi \psi$, we have $\overline{\lambda} = \oplus_{\psi \in \widehat{G/N}} d_\psi \overline{\psi}$, from which (1.73) immediately follows.

Applying Theorem 1.3.2(ii) and recalling that $I_G(\sigma) = G$ we find that

$$\operatorname{Hom}_G(\operatorname{Ind}_N^G \sigma, \operatorname{Ind}_N^G \sigma) = \mathbb{C}^{|G/N|}.$$

Since $|G/N| = \sum_{\psi \in \widehat{G/N}} d_\psi^2$ (see again [11, Theorem 3.7.11(iii)]), from (1.73) and the commutant theorem (Theorem 1.2.2) we obtain

$$|G/N| = \dim[\mathrm{Hom}_G(\mathrm{Ind}_N^G \sigma, \mathrm{Ind}_N^G \sigma)]$$
$$= \sum_{\psi_1, \psi_2 \in \widehat{G/N}} d_{\psi_1} d_{\psi_2} \dim[\mathrm{Hom}_G(\theta \otimes \overline{\psi_1}, \theta \otimes \overline{\psi_2})],$$

which implies the irreducibility and pairwise inequivalence of the $\theta \otimes \overline{\psi}$'s as well as the equality of d_ψ with the inertia index of $\theta \otimes \overline{\psi}$ with respect to N (by Frobenius reciprocity). Thus (i) and (ii) are proved.

Suppose now that $\tau \in \widehat{G}$ and $\mathrm{Res}_N^G \tau = \sigma$. From Corollary 1.1.16 and (1.73) we deduce that

$$\tau \otimes \overline{\lambda} \sim \mathrm{Ind}_N^G \mathrm{Res}_N^G \tau \sim \mathrm{Ind}_N^G \sigma = \bigoplus_{\psi \in \widehat{G/N}} d_\psi (\theta \otimes \overline{\psi}),$$

so that, observing that $\dim \mathrm{Hom}_G(\iota_G, \overline{\lambda}) = 1$ (which follows immediately from Frobenius reciprocity), we necessarily have $\tau \sim \tau \otimes \iota_G \sim \theta \otimes \overline{\psi}$ for some $\psi \in \widehat{G/N}$ and $d_\psi = 1$. $\square$

1.3.2 The little group method

In this subsection we give a general formulation of the little group method. As an example, we apply it to semidirect products with an Abelian normal subgroup. This particular case was developed by Frobenius (for finite groups) and then by Wigner [76] for groups arising in physical problems as the Poincaré group (see [73, Sections 3.9 and 3.20]). Finally, Mackey extended this method to topological groups [55] (see also [56]).

Definition 1.3.10 Let G be a group, $H \leq G$ a subgroup of G and $\sigma \in \widehat{H}$. An *extension* of σ to G is a representation $\widetilde{\sigma} \in \widehat{G}$ such that $\mathrm{Res}_H^G \widetilde{\sigma} = \sigma$.

Theorem 1.3.11 (The little group method) *Let G be a finite group and $N \trianglelefteq G$ a normal subgroup. Suppose that any $\sigma \in \widehat{N}$ has an extension $\widetilde{\sigma}$ to its inertia group $I_G(\sigma)$ (see Definition 1.3.1). In $\widehat{N}$ define an equivalence relation $\approx$ by setting $\sigma_1 \approx \sigma_2$ if there exists $g \in G$ such that $^g\sigma_1 \sim \sigma_2$. Let Σ be a set of representatives of the corresponding quotient space $\widehat{N}/\approx$.*

For $\psi \in \widehat{I_G(\sigma)/N}$ denote by $\overline{\psi} \in \widehat{I_G(\sigma)}$ its inflation to $I_G(\sigma)$ (see (1.71)). Then

$$\widehat{G} = \{\mathrm{Ind}_{I_G(\sigma)}^G (\widetilde{\sigma} \otimes \overline{\psi}) : \sigma \in \Sigma, \psi \in \widehat{I_G(\sigma)/N}\},$$

that is, the right-hand side is a list of all irreducible G-representations where for different values of σ and ψ we obtain inequivalent representations.

Proof Let $\sigma \in \Sigma$. From Theorem 1.3.9 (with $I_G(\sigma)$ in place of G and $\tilde{\sigma}$ in place of θ) we deduce that

$$\text{Ind}_N^{I_G(\sigma)} \sigma = \bigoplus_{\psi \in \widehat{I_G(\sigma)/N}} d_\psi (\tilde{\sigma} \otimes \overline{\psi}) \tag{1.74}$$

where the $\tilde{\sigma} \otimes \overline{\psi}$ are irreducible and pairwise inequivalent representations in $\widehat{I_G(\sigma)}$. From the Clifford correspondence (Theorem 1.3.6) we find that the G-representations

$$\text{Ind}_{I_G(\sigma)}^G (\tilde{\sigma} \otimes \overline{\psi}), \tag{1.75}$$

$\psi \in \widehat{I_G(\sigma)}$, are irreducible and pairwise inequivalent. By Theorem 1.3.2(i), the restriction to N of a representation as in (1.75) (which is contained in $\text{Ind}_N^G \sigma$) is a sum of G-conjugates of σ. It follows that the representations corresponding to $(\sigma_1, \psi_1) \neq (\sigma_2, \psi_2)$ $(\sigma_i \in \Sigma, \psi_i \in \widehat{I_G(\sigma_i)}, i = 1, 2)$ are inequivalent (recall that Σ is a system of representatives for the orbits of G on $\widehat{N}$).

Finally, if τ is an irreducible representation of G, by Theorem 1.3.2(iii) we can find a $\sigma \in \Sigma$ such that $\tau \in \widehat{G}(\sigma)$, and therefore, again by Clifford correspondence, τ is of the form $\text{Ind}_{I_G(\sigma)}^G \xi$ with ξ an irreducible representation of $I_G(\sigma)$ such that $\sigma \preceq \text{Res}_N^{I_G(\sigma)} \xi$. But then $\xi \preceq \text{Ind}_N^{I_G(\sigma)} \sigma$, and therefore by (1.74) there exists $\psi \in \widehat{I_G(\sigma)/N}$ such that $\xi = \tilde{\sigma} \otimes \overline{\psi}$. $\qquad\square$

1.3.3 Semidirect products

In this section, we recall a well known construction in group theory (see, for instance, [2, pp. 20–24] or [68, pp. 6–8]).

Definition 1.3.12 (Semidirect product) Let G be a finite group and $N, H \leq G$ two subgroups of G. Then G is the (internal) *semidirect product* of N by H, and we write $G = N \rtimes H$ when the following conditions are satisfied:

(i) $N \trianglelefteq G$;

(ii) $G = NH$;

(iii) $N \cap H = \{1_G\}$.

Proposition 1.3.13 *Suppose that G is a semidirect product of N by H. Then:*

(i) *$G/N \cong H$;*

(ii) *every $g \in G$ has a unique expression of the form $g = nh$ with $n \in N$ and $h \in H$;*

(iii) *for any* $h \in H$ *and* $n \in N$ *set* $\phi_h(n) = hnh^{-1}$. *Then* $\phi_h \in \mathrm{Aut}(N)$ *for all* $h \in H$, *and the map*

$$
\begin{array}{ccc}
H & \longrightarrow & \mathrm{Aut}(N) \\
h & \longmapsto & \phi_h
\end{array}
$$

is a homomorphism (the conjugation homomorphism*);*

(iv) *if* $nh, n_1 h_1 \in G$ *as in (ii) then their product is given by*

$$(n_1 h_1)(n_2 h_2) = [n_1 h_1 n_2 h_1^{-1}]h_1 h_2 = [n_1 \phi_{h_1}(n_2)]h_1 h_2. \tag{1.76}$$

Conversely, suppose that H *and* N *are two (finite) groups with a homomorphism* $H \ni h \mapsto \phi_h \in \mathrm{Aut}(N)$. *Set* $G = \{(n, h) : n \in N, h \in H\}$ *and define a product in* G *by setting*

$$(n, h)(n_1, h_1) = (n\phi_h(n_1), hh_1)$$

(see (1.76)). Then G *is a group and is isomorphic to the (inner) semidirect product of* $\widetilde{N} = \{(n, 1_H) : n \in N\} \cong N$ *by* $\widetilde{H} = \{(1_N, h) : H \in H\} \cong H$. *The group* G *is called the* external semidirect product *of* N *by* H *with respect to* ϕ *and is usually denoted by* $N \rtimes_\phi H$.

Moreover, with the above notation, the following conditions are equivalent:

- G *is isomorphic to the direct product* $\widetilde{N} \times \widetilde{H}$;
- $\widetilde{H}$ *is normal in* G;
- ϕ_h *is the trivial automorphism of* N *for all* $h \in H$.

Proof The proof is an easy exercise and is left to the reader. □

Clearly, the internal and external semidirect products are equivalent constructions and we shall make no distinction between them.

1.3.4 Semidirect products with an Abelian normal subgroup

We now apply the little group method to an important class of semidirect products, namely that of semidirect products with an Abelian normal subgroup. The approach with our version of the little group method considerably simplifies the setting.

In the following, we adopt the convention of identifying any irreducible representation σ of an Abelian group A with its character $\chi = \chi_\sigma$. Recall that $\chi(a^{-1}) = \overline{\chi(a)}$ (the complex conjugate) for all $a \in A$.

Theorem 1.3.14 *Suppose that* $G = A \rtimes H$ *with* A *Abelian. Given* $\chi \in \widehat{A}$, *its inertia group* $I_G(\chi)$ *coincides with* $A \rtimes H_\chi$ *where* $H_\chi = \{h \in H : {}^h\chi = \chi\}$.

Then any $\chi \in \widehat{A}$ *may be extended to a one-dimensional representation* $\widetilde{\chi} \in \widehat{A \rtimes H_\chi}$ *by setting*

$$\widetilde{\chi}(ah) = \chi(a) \qquad \forall a \in A, \ h \in H_\chi. \tag{1.77}$$

Moreover, with the notation used in Theorem 1.3.11,

$$\widehat{G} = \{\mathrm{Ind}_{A \rtimes H_\chi}^G (\widetilde{\chi} \otimes \overline{\psi}) : \chi \in \Sigma, \ \psi \in \widehat{H_\chi}\}.$$

Proof For $a, a_1 \in A$ and $h \in H$ we have

$$\begin{aligned}
{}^{ah}\chi(a_1) &= \chi(h^{-1}a^{-1}a_1ah) \\
&= \chi(h^{-1}a^{-1}h)\chi(h^{-1}a_1h)\chi(h^{-1}ah) \\
&= \chi(h^{-1}a_1h) = {}^h\chi(a_1),
\end{aligned}$$

thus showing that the inertia subgroup of χ coincides with $A \rtimes H_\chi$.

Let $\chi \in \widehat{A}$ and let us prove that the extension of χ defined by (1.77) is a homomorphism. By the definition of H_χ we have that χ is invariant under conjugation with elements in H_χ, so that if $a_1, a_2 \in A$ and $h_1, h_2 \in H_\chi$ we have

$$\begin{aligned}
\widetilde{\chi}((a_1h_1)(a_2h_2)) &= \widetilde{\chi}(a_1h_1a_2h_1^{-1}(h_1h_2)) \\
&= \chi(a_1h_1a_2h_1^{-1}) \\
&= \chi(a_1)\chi(a_2) \\
&= \widetilde{\chi}(a_1h_1)\widetilde{\chi}(a_2h_2).
\end{aligned}$$

Finally, the last statement is just an application of Theorem 1.3.11. $\qquad\square$

1.3.5 The affine group over a finite field

This subsection is based on Chapters 16 and 17 of Terras' monograph [74]. See also Diaconis' book [20, Chapter 3, Section D, Example 4].

Let $\mathbb{F}_q$ be a finite field, where $q = p^r$ with p a prime number and r a positive integer. We denote by $\mathbb{F}_q^* = \{x \in \mathbb{F}_q : x \neq 0\}$ the multiplicative group of invertible elements in $\mathbb{F}_q$.

The *affine group over* $\mathbb{F}_q$ is the group of matrices

$$\mathrm{Aff}(\mathbb{F}_q) = \left\{ \begin{pmatrix} x & y \\ 0 & 1 \end{pmatrix} : x \in \mathbb{F}_q^*, y \in \mathbb{F}_q \right\}.$$

It acts on $\mathbb{F}_q \equiv \left\{ \begin{pmatrix} t \\ 1 \end{pmatrix} : t \in \mathbb{F}_q \right\}$ by multiplication:

$$\begin{pmatrix} x & y \\ 0 & 1 \end{pmatrix} \begin{pmatrix} t \\ 1 \end{pmatrix} = \begin{pmatrix} xt + y \\ 1 \end{pmatrix}.$$

Therefore, we may also consider $\mathrm{Aff}(\mathbb{F}_q)$ as the group of affine transformations of $\mathbb{F}_q$:

$$\mathrm{Aff}(\mathbb{F}_q) \cong \{\tau_{x,y} : x \in \mathbb{F}_q^*, y \in \mathbb{F}_q\},$$

where $\tau_{x,y}(t) = xt + y$ for all $t \in \mathbb{F}_q$. Note that

$$\tau_{x,y}\tau_{u,v} = \tau_{xu,xv+y}.$$

In order to simplify the notation, we shall identify $\mathrm{Aff}(\mathbb{F}_q)$ with the set $\{(x, y) : x \in \mathbb{F}_q^*, y \in \mathbb{F}_q\}$ equipped with composition law

$$(x, y)(u, v) = (xu, xv + y). \tag{1.78}$$

Remark 1.3.15   In the lemma below we shall prove that $\mathrm{Aff}(\mathbb{F}_q)$ is a semi-direct product. However, the notation in (1.78) differs from the standard notation (see Proposition 1.3.13) since the coordinates of the elements of the corresponding direct product are switched.

Lemma 1.3.16   (i) *The identity of* $\mathrm{Aff}(\mathbb{F}_q)$ *is* $(1, 0)$ *and the inverse of* (x, y) *is* $(x, y)^{-1} = (x^{-1}, -x^{-1}y)$.

(ii) *Setting* $A = \{(1, y) : y \in \mathbb{F}_q\} \cong \mathbb{F}_q$ *(a group with respect to addition) and* $H = \{(x, 0) : x \in \mathbb{F}_q^*\} \cong \mathbb{F}_q^*$ *(a group with respect to multiplication), we have*

$$\mathrm{Aff}(\mathbb{F}_q) \cong A \rtimes H$$

and the corresponding homomorphism $\mathbb{F}_q^* \ni x \mapsto \phi_x \in \mathrm{Aut}(\mathbb{F}_q)$ *is given by* $\phi_x(y) = xy$ *for all* $x \in \mathbb{F}_q^*, y \in \mathbb{F}_q$.

(iii) *The conjugacy classes of the group* $\mathrm{Aff}(\mathbb{F}_q)$ *are* $C_0 = \{(1, 0)\}, C_1 = \{(1, y) : y \in \mathbb{F}_q^*\}$ *and* $C_x = \{(x, y) : y \in \mathbb{F}_q\}, x \in \mathbb{F}_q^*, x \neq 1$.

Proof   Part (i) is trivial, while (ii) and (iii) are immediate consequences of the identity

$$(u, v)(x, y)(u, v)^{-1} = (x, -xv + uy + v) \equiv (x, (1 - x)v + uy).$$

$\square$

Since $\mathrm{Aff}(\mathbb{F}_q)$ is a semidirect product with an abelian normal subgroup, we can apply Theorem 1.3.11 (the little group method) to get a list of all irreducible representations of $\mathrm{Aff}(\mathbb{F}_q)$. As usual, $\widehat{\mathbb{F}_q}$ (resp. $\widehat{\mathbb{F}_q^*}$) will denote the dual of the additive group $\mathbb{F}_q$ (resp. the multiplicative group $\mathbb{F}_q^*$). From Lemma 1.3.16(ii) it follows that the conjugacy action of $\mathbb{F}_q^*$ on $\widehat{\mathbb{F}_q}$ is given by

$$^a\chi(x) = \chi(a^{-1}x) \tag{1.79}$$

for all $\chi \in \widehat{\mathbb{F}_q}, x \in \mathbb{F}_q, a \in \mathbb{F}_q^*$. Denote by $\chi_0 \equiv 1$ the trivial character of $\mathbb{F}_q$.

Lemma 1.3.17   *The action of $\mathbb{F}_q^*$ on $\widehat{\mathbb{F}_q}$ has exactly two orbits, namely $\{\chi_0\}$ and $\widehat{\mathbb{F}_q} \setminus \{\chi_0\}$. Moreover, the stabilizer of $\chi \in \widehat{\mathbb{F}_q}$ is given by*

$$(\mathbb{F}_q^*)_\chi = \begin{cases} \{1_{\mathbb{F}_q^*}\} & \text{if } \chi \neq \chi_0 \\ \mathbb{F}_q^* & \text{if } \chi = \chi_0. \end{cases}$$

Proof   It is clear that χ_0 is a fixed point. From now on, let $\chi \in \widehat{\mathbb{F}_q}$ be a nontrivial character. For $a \in \mathbb{F}_q$ let us set

$$^a\chi^* = \begin{cases} a^{-1}\chi & \text{if } a \in \mathbb{F}_q^* \\ \chi_0 & \text{if } a = 0. \end{cases}$$

We claim that the map $a \mapsto {}^a\chi^*$ yields an isomorphism from $\mathbb{F}_q$ onto $\widehat{\mathbb{F}_q}$. Indeed it is straightforward to check that $^{(a+b)}\chi^*(x) = {}^a\chi^*(x) {}^b\chi^*(x)$ for all $a, b, x \in \mathbb{F}_q$. Moreover, if $a \neq 0$ we have $^a\chi^* \neq \chi_0$ since the map $x \mapsto ax$ is a bijection of $\mathbb{F}_q$. This shows that the homomorphism $a \mapsto {}^a\chi^*$ is injective. Since $|\mathbb{F}_q| = |\widehat{\mathbb{F}_q}|$, it is in fact bijective. As a consequence, we have that $\{^a\chi : a \neq 0\} = \{^a\chi^* : a \neq 0\}$ coincides with the set of nontrivial characters. $\square$

Theorem 1.3.18   *The group $\mathrm{Aff}(\mathbb{F}_q)$ has exactly $q-1$ one-dimensional representations and one $(q-1)$-dimensional irreducible representation. The former are obtained by associating with each $\psi \in \widehat{\mathbb{F}_q^*}$ the function $\Psi \colon \mathrm{Aff}(\mathbb{F}_q) \to \mathbb{C}$ defined by*

$$\Psi(x, y) = \psi(x)$$

for all $(x, y) \in \mathrm{Aff}(\mathbb{F}_q)$. The $(q-1)$-dimensional irreducible representation is given by

$$\mathrm{Ind}_{\mathbb{F}_q}^{\mathrm{Aff}(\mathbb{F}_q)} \chi, \tag{1.80}$$

where χ is any nontrivial character of $\mathbb{F}_q$.

Proof   This is just an application of the little group method (Theorem 1.3.11). Indeed, the inertia group of the trivial character $\chi_0 \in \widehat{\mathbb{F}_q}$ is $\mathrm{Aff}(\mathbb{F}_q)$, by Lemma 1.3.17. This provides the $q-1$ one-dimensional representations simply by taking any character $\psi \in \widehat{\mathbb{F}_q^*}$. The inertia group of a nontrivial character χ is $\mathbb{F}_q$ since, by Lemma 1.3.17, $(\mathbb{F}_q^*)_\chi = \{1_{\mathbb{F}_q^*}\}$. $\square$

Exercise 1.3.19   Show that $\left(\mathrm{Aff}(\mathbb{F}_q), \mathbb{F}_q^*\right)$ is a symmetric Gelfand pair with homogeneous space $\mathrm{Aff}(\mathbb{F}_q)/\mathbb{F}_q^* \equiv \mathbb{F}_q$ and that the two spherical representations are the trivial representation and the one in (1.80), with corresponding spherical functions $\phi_0 \equiv 1$ and ϕ_1 given by

$$\phi_1(x) = \begin{cases} 1 & \text{if } x = 0 \\ -\frac{1}{q-1} & \text{if } x \neq 0 \end{cases}$$

for all $x \in \mathbb{F}_q$.

Hint. The group $\text{Aff}(\mathbb{F}_q)$ acts on $\mathbb{F}_q$ doubly transitively and one may apply Exercise 1.2.34.

We now list some more advanced topics on the structure of finite fields that shed more light on the representation theory of $\text{Aff}(\mathbb{F}_q)$. We refer to the books by Lang [49, Chapter 6] and Winnie Li [53, Chapter 1] for complete proofs and further details.

Theorem 1.3.20 *Let $\mathbb{F}_q$ be the finite field with $q = p^r$ elements and p a prime number.*

(i) *The multiplicative group $\mathbb{F}_q^*$ is cyclic (of order $q - 1$); in other words there exists $g \in \mathbb{F}_q^*$ such that $\mathbb{F}_q^* = \{1, g, g^2, \ldots, g^{q-2}\}$.*

(ii) *The additive group of $\mathbb{F}_q$ is isomophic to the direct sum of r copies of the (additive) cyclic group of order p.*

(iii) *If we set $\text{tr}(x) = x + x^p + x^{p^2} + \cdots + x^{p^{r-1}}$ then tr (called the trace) is a homomorphism of the additive group $\mathbb{F}_q$ onto the additive group $\mathbb{F}_p$. Moreover the kernel of tr is given by the subgroup $\text{Ker}(\text{tr}) = \{y - y^p : y \in \mathbb{F}_q\}$ (Hilbert's Satz 90).*

From Theorem 1.3.20(iii) we deduce that the expression

$$\chi(x) = \exp[2\pi i \, \text{tr}(x)/p], \qquad x \in \mathbb{F}_q \tag{1.81}$$

(we identify $\mathbb{F}_p$ with $\{0, 1, \ldots, p - 1\}$), defines a nontrivial character $\chi \in \widehat{\mathbb{F}_q}$. Recall (see Theorem 1.3.18) that we may use this character to get the (unique) higher-dimensional irreducible representation (1.80) of $\text{Aff}(\mathbb{F}_q)$.

Exercise 1.3.21 Show that $\widehat{\mathbb{F}_q} = \{\chi_s : s \in \mathbb{F}_q\}$, where

$$\chi_s(x) = \chi(sx) = \exp[2\pi i \, \text{tr}(sx)/p] \tag{1.82}$$

for all $s, r \in \mathbb{F}_q$.
   Deduce that $\widehat{\mathbb{F}_q^2} = \{\chi_{s,t} : s, t \in \mathbb{F}_q\}$, where

$$\chi_{s,t}(x, y) = \chi(sx + ty) = \exp[2\pi i \, \text{tr}(sx + ty)/p] \tag{1.83}$$

for all $s, t, x, y \in \mathbb{F}_q$.

Exercise 1.3.22   Use (1.81) together with Theorem 1.3.20(ii) to show that a matrix realization of $\operatorname{Ind}_{\mathbb{F}_q}^{\operatorname{Aff}(\mathbb{F}_q)} \chi$ is given by

$$U(g^k, y) = D(y)W^k,$$

where g is a cyclic generator of $\mathbb{F}_q^*$, $k = 0, 1, \ldots, q - 2$, $y \in \mathbb{F}_q$, $D(y)$ is the $(q - 1) \times (q - 1)$ diagonal matrix

$$D(y) = \begin{pmatrix} \chi(y) & 0 & 0 & \cdots & 0 \\ 0 & \chi(gy) & 0 & \cdots & 0 \\ \vdots & \vdots & \vdots & \ddots & \vdots \\ 0 & 0 & 0 & \cdots & \chi(g^{q-2}y) \end{pmatrix}$$

and W is the $(q - 1) \times (q - 1)$ permutation matrix

$$W = \begin{pmatrix} 0 & 1 & 0 & \cdots & 0 \\ 0 & 0 & 1 & \cdots & 0 \\ \vdots & \vdots & \vdots & \ddots & \vdots \\ 0 & 0 & 0 & \cdots & 1 \\ 1 & 0 & 0 & \cdots & 0 \end{pmatrix}.$$

Hint. Use equation (1.7) with $S = \{g^{-i} : i = 0, 1, \ldots, q - 2\}$ as a system of representatives for the left cosets of $\mathbb{F}_q$ in $\operatorname{Aff}(\mathbb{F}_q)$. Use the identities

$$(g^k, y) = (1, y)(g^k, 0) = (1, y)(g, 0)^k,$$
$$(g^i, 0)(1, x)(g^{-j}, 0) = (g^{i-j}, g^i x),$$
$$(g^i, 0)(g, 0)(g^{-j}, 0) = (g^{i-j+1}, 0)$$

for all $i, j, k = 0, 1, \ldots, q - 2$.

1.3.6 The finite Heisenberg group

This subsection, which is a natural continuation of the preceding one, is based on Chapter 18 of Terras' monograph [74].

Let $\mathbb{F}_q$ be a finite field and $q = p^r$ with p a prime number. The *Heisenberg group* over $\mathbb{F}_q$ is the group of matrices

$$\mathcal{H}_q = \left\{ \begin{pmatrix} 1 & x & z \\ 0 & 1 & y \\ 0 & 0 & 1 \end{pmatrix} : x, y, z \in \mathbb{F}_q \right\}.$$

Exercise 1.3.23

(i) Show that $\mathcal{H}_q$ is isomorphic to the set $\{(x, y, z) : x, y, z \in \mathbb{F}_q\}$ endowed with the composition law

$$(x, y, z)(u, v, w) = (x + u, y + v, xv + w + z). \qquad (1.84)$$

In particular, check that

$$(x, y, z)^{-1} = (-x, -y, -z + xy).$$

(ii) Deduce from (1.84) that

$$(x, y, z) = (0, 0, z)(0, y, 0)(x, 0, 0). \qquad (1.85)$$

(iii) Show that

$$(x, y, z)^{-1}(u, v, w)(x, y, z) = (u, v, uy - xv + w) \qquad (1.86)$$

and deduce that the conjugacy classes of $\mathcal{H}_q$ are $C_w = \{(0, 0, w)\}$, $w \in \mathbb{F}_q$ (giving q one-element classes) and $C_{u,v} = \{(u, v, w) : w \in \mathbb{F}_q\}$, $u, v \in \mathbb{F}_q$, $(u, v) \neq (0, 0)$ (giving $q^2 - 1$ classes of q elements).

(iv) From (1.86) deduce that $(x, 0, 0)^{-1}(0, v, w)(x, 0, 0) = (0, v, w - xv)$ and thus, in turn, that

$$\mathcal{H}_q \cong \mathbb{F}_q^2 \rtimes_\phi \mathbb{F}_q,$$

where $\mathbb{F}_q^2 = \{(0, v, w) : v, w \in \mathbb{F}_q\}$ and $\mathbb{F}_q = \{(x, 0, 0) : x \in \mathbb{F}_q\}$ are viewed as additive groups and ϕ is the $\mathbb{F}_q$-action on $\mathbb{F}_q^2$ given by

$$\phi_x(v, w) = (v, w - xv)$$

with $x, y, z \in \mathbb{F}_q$.

As a consequence of Exercise 1.3.23(iii), we shall denote the elements of $\mathcal{H}_q$ by $(x, y, z) \in \mathbb{F}_q \times \mathbb{F}_q^2$ with multiplication as in (1.84).

Using the notation from Theorem 1.3.14 (with $G = \mathcal{H}_q$, $A = \mathbb{F}_q^2$ and $H = \mathbb{F}_q$), given $\chi_{s,t} \in \widehat{\mathbb{F}_q^2}$ (see (1.83)), we have

$$H_{\chi_{s,t}} = \begin{cases} \{1_H\} & \text{if } t \neq 0 \\ H & \text{otherwise.} \end{cases}$$

Indeed, from

$$\begin{aligned}
{}^{(x,0,0)}\chi_{s,t}(v, w) &= \chi_{s,t}(v, w - xv) \\
&= \chi(sv + t(w - xv)) \\
&= \chi((s - tx)v + tw) \\
&= \chi_{s-tx,t}(v, w)
\end{aligned}$$

we deduce that $^{(x,0,0)}\chi_{s,t} = \chi_{s,t}$ if and only if either $t = 0$ (in this case, the $\approx$ equivalence class of each $\chi_{s,0}$ reduces to the element $\chi_{s,0}$ itself and therefore $H_{\chi_{s,0}} = H$) or $t \neq 0$ and $x = 0$ (so that $H_{\chi_{s,t}} = \{1_H\}$).

According to the preceding analysis, we can choose

$$\Sigma = \{\chi_{s,0} : s \in \mathbb{F}_q\} \cup \{\chi_{0,t} : t \in \mathbb{F}_q, t \neq 0\}$$

as a set of representatives of the quotient space $\widehat{A}/\approx$ (cf. Theorem 1.3.11).

For every $s, u \in \mathbb{F}_q$ if we denote by $\psi_{s,u} \in \widehat{\mathcal{H}_q}$ the character defined by

$$\psi_{s,u}(x, y, z) = \chi(sy + ux)$$

then, recalling that $H_{\chi_{s,0}} = H$ (so that $A \rtimes H_{\chi_{s,0}} = G$) and that $\overline{\chi_u} \in \widehat{G}$ denotes the inflation of $\chi_u \in \widehat{G/A} = \widehat{H} = \widehat{\mathbb{F}_q}$, we have

$$\text{Ind}_{A \rtimes H_{\chi_{s,0}}}^{G} (\widetilde{\chi_{s,0}} \otimes \overline{\chi_u})(x, y, z) = (\widetilde{\chi_{s,0}} \otimes \overline{\chi_u})(x, y, z)$$

$$= \chi_{s,0}(y, z)\chi_u(x)$$

$$= \chi(sy + ux)$$

$$= \psi_{s,u}(x, y, z)$$

so that

$$\text{Ind}_{A \rtimes H_{\chi_{s,0}}}^{G} (\widetilde{\chi_{s,0}} \otimes \overline{\chi_u}) = \psi_{s,u}.$$

Moreover, if $t \neq 0$ then $H_{\chi_{0,t}} = \{1_H\}$ (so that $A \rtimes H_{\chi_{0,t}} = A$), and setting $\pi_t := \text{Ind}_A^G(\widetilde{\chi_{0,t}}) \in \widehat{\mathcal{H}_q}$ we have

$$\text{Ind}_{A \rtimes H_{\chi_{0,t}}}^{G} (\widetilde{\chi_{0,t}}) = \text{Ind}_A^G(\widetilde{\chi_{0,t}}) = \pi_t. \tag{1.87}$$

From Theorem 1.3.14 we deduce that $\widehat{\mathcal{H}_q}$ consists exactly of the q^2 one-dimensional representations $\psi_{s,u}$, $s, u \in \mathbb{F}_q$, and the $q - 1$ representations π_t, $t \in \mathbb{F}_q, t \neq 0$, of dimension $[G : A] = |H| = |\mathbb{F}_q| = q$.

Exercise 1.3.24 Use (1.87) to show that a matrix realization of π_t is given by

$$U(x, y, z) = \chi(tz)D(ty)W(x),$$

where $D(ty)$ is the $q \times q$ diagonal matrix

$$D(ty) = \begin{pmatrix} \chi(0) & 0 & 0 & 0 & \cdots & 0 \\ 0 & \chi(ty) & 0 & 0 & \cdots & 0 \\ 0 & 0 & \chi(2ty) & 0 & \cdots & 0 \\ \vdots & \vdots & \vdots & \vdots & \ddots & \vdots \\ 0 & 0 & 0 & 0 & \cdots & \chi((q-1)ty) \end{pmatrix}$$

and $W(x)$ is the $q \times q$ permutation matrix defined by

$$W(x)_{i,j} = \delta_i(j + x)$$

for all $i, j \in \mathbb{F}_q$.

Hint. Use equation (1.85) and observe that $S = \{(i, 0, 0) : i \in \mathbb{F}_q\} = H = \mathbb{F}_q$ is a system of representatives for the left cosets of $A = \mathbb{F}_q^2$ in $G = \mathcal{H}_q$. Use the identities

$$(i, 0, 0)(0, 0, z)(-j, 0, 0) = (i - j, 0, z),$$

$$(i, 0, 0)(0, y, 0)(-j, 0, 0) = (i - j, y, iy),$$

$$(i, 0, 0)(x, 0, 0)(-j, 0, 0) = (i - j + x, 0, 0)$$

for all $i, j, x, y, z \in \mathbb{F}_q$.

2

Wreath products of finite groups and their representation theory

In this chapter, which constitutes the core of the book, we develop the representation theory of wreath products. Our exposition is inspired by the monographs of James and Kerber [38] and Huppert [35]. Howewer, our approach is more analytical and, in particular, we interpret the exponentiation and the composition actions in terms of actions on suitable rooted trees. This is done in Section 2.1.2. In Section 2.3 we describe the conjugacy classes of wreath products $F \wr G$, with particular emphasis on groups of the form $C_2 \wr G$ (Section 2.3.2), and $F \wr S_n$ (Section 2.3.3), and then in Section 2.4 we use the little group method (Theorem 1.3.11) to determine a complete list of irreducible representations of wreath products. Finally in Sections 2.5 and 2.6 we analyze the representation theory of groups of the form $C_2 \wr G$ and $F \wr S_n$, respectively. This yields, in particular, a clear description of the representations of finite lamplighter groups (Sections 2.5.1 and 2.5.2) as well as of the groups $S_m \wr S_n$ (Section 2.6.1).

2.1 Basic properties of wreath products of finite groups

2.1.1 Definitions

Let G and F be two finite groups and suppose that G acts on a finite set X. Denote by F^X the set of all maps $f : X \to F$. The set F^X is a group under pointwise multiplication: $(f \cdot f')(x) = f(x)f'(x)$ for all $f, f' \in F^X$ and $x \in X$. We can define a natural action of G on F^X by setting $(gf)(x) = f(g^{-1}x)$ for all $g \in G$, $f \in F^X$ and $x \in X$. We have $g(f \cdot f') = gf \cdot gf'$ and $(gf)^{-1} = gf^{-1}$; in this way G acts on F^X as a group of automorphisms.

Define a multiplication on the set $F^X \times G = \{(f, g) : f \in F^X, g \in G\}$ by setting

$$(f, g)(f', g') = (f \cdot gf', gg') \qquad (2.1)$$

for all $(f, g), (f', g') \in F^X \times G$, where, with the above notation,

$$(f \cdot gf')(x) = f(x)f'(g^{-1}x) \qquad (2.2)$$

for all $x \in X$.

Lemma 2.1.1   *The set $F^X \times G$ equipped with the multiplication (2.1) is a group. The identity element is $(1_F, 1_G)$, where $1_F(x) = 1_F$ for all $x \in X$, and the inverse of (f, g) is given by $(g^{-1}f^{-1}, g^{-1})$.*

Proof   Let $(f, g), (f', g'), (f'', g'') \in F^X \times G$. Then, we have

$$\begin{aligned}
[(f, g)(f', g')](f'', g'') &= (f \cdot gf', gg')(f'', g'') \\
&= ((f \cdot gf') \cdot gg'f'', (gg')g'') \\
&= (f \cdot (gf' \cdot gg'f''), g(g'g'')) \\
&= (f \cdot g(f' \cdot g'f''), g(g'g'')) \\
&= (f, g)(f' \cdot g'f'', g'g'') \\
&= (f, g)[(f', g')(f'', g'')].
\end{aligned}$$

This shows that the operation (2.1) is associative. It easy to show that $(1_F, 1_G)$ is the identity. Moreover, we have

$$\begin{aligned}
(g^{-1}f^{-1}, g^{-1})(f, g) &= (g^{-1}f^{-1} \cdot g^{-1}f, 1_G) \\
&= (g^{-1}(f^{-1} \cdot f), 1_G) \\
&= (1_F, 1_G) \\
&= (f, g)(g^{-1}f^{-1}, g^{-1})
\end{aligned}$$

and therefore $(g^{-1}f^{-1}, g^{-1})$ is the inverse of (f, g). $\square$

Definition 2.1.2

(i) The set $F^X \times G$ when equipped with the above group structure is called the *wreath product* of F by the permutation group G and is denoted by $F \wr G$.

(ii) The subgroup

$$\overline{F^X} = \{(f, 1_G) : f \in F^X\}$$

is called the *base group*: it is naturally identified with F^X.

(iii) The subgroup $\mathrm{diag}F^X := \{f \in F^X : f \text{ is constant on } X\} \cong F$ is called the *diagonal subgroup* of the base group.

It is easy to show that the base group is a normal subgroup of $F \wr G$. Moreover, the wreath product may be written as the semidirect product (cf. Definition 1.3.12) of the base group by the subgroup $\overline{G} = \{(\mathbf{1}_F, g) : g \in G\} \cong G$ (note that $(f, g) = (f, 1_G) \cdot (\mathbf{1}_F, g)$ for all $(f, g) \in F \wr G$). Modulo the identification of $\overline{G}$ with G we thus have

$$F \wr G = F^X \rtimes G.$$

Since

$$(\mathbf{1}_F, g)(f, 1_G) = (gf, g) = (f, g) = (f, 1_G)(\mathbf{1}_F, g)$$

for all $g \in G$ and $f \in \mathrm{diag}\, F^X$, and since $(\mathrm{diag}\, F^X) \cap \overline{G} = \{(\mathbf{1}_F, 1_G)\}$, we deduce that $(\mathrm{diag}\, F^X)\overline{G}$, as a subgroup of $F \wr G$, is isomorphic to the direct product $F \times G$.

Proposition 2.1.3 *Let G (resp. G_1, resp. G_2) be a finite group acting on a finite set X (resp. X_1, resp. X_2). Also let F be a finite group and $H \leq G$ a subgroup. Then*

$$(F \wr G)/(F \wr H) \cong G/H \tag{2.3}$$

and

$$F \wr (G_1 \times G_2) \cong (F \wr G_1) \times (F \wr G_2) \tag{2.4}$$

(here $G_1 \times G_2$ acts on $X_1 \coprod X_2$ as follows: $(g_1, g_2)(x_i) = g_i x_i$ for $x_i \in X_i$, $i = 1, 2$).

Proof For $g_1, g_2 \in G$ we write $g_1 \sim_H g_2$ if there exists $h \in H$ such that $g_1 = hg_2$ or equivalently if g_1 and g_2 belong to the same right H-coset: $Hg_1 = Hg_2$. Analogously, for $f_1, f_2 \in F^X$ and $g_1, g_2 \in G$ we write $(f_1, g_2) \sim_{F \wr H} (f_2, g_2)$ if there exists $(f, h) \in F \wr H$ such that $(f_1, g_2) = (f, h)(f_2, g_2)$. Denoting as usual by $\mathbf{1}_F \in F^X$ the constant function $\mathbf{1}_F(x) = 1_F$, where 1_F is the identity element in F, we have $(f, g) = (f, 1_H)(\mathbf{1}_F, g)$ for all $f \in F^X$ and $g \in G$. This shows that $(f, g) \sim_{F \wr H} (\mathbf{1}_F, g)$ for all $f \in F^X$ and $g \in G$. Also, if $g_1, g_2 \in G$ and $g_1 \sim_H g_2$, say $g_1 = hg_2$ for $h \in H$, then $(\mathbf{1}_F, g_1) = (\mathbf{1}_F, h)(\mathbf{1}_F, g_2)$ and therefore $(\mathbf{1}_F, g_1) \sim_{F \wr H} (\mathbf{1}_F, g_2)$. Vice versa, if $(\mathbf{1}_F, g_1) \sim_{F \wr H} (\mathbf{1}_F, g_2)$ then one easily shows that $g_1 \sim_H g_2$. This proves (2.3).

Let us set $X = X_1 \coprod X_2$. We leave it to the reader to check that the map

$$\begin{aligned} F \wr (G_1 \times G_2) &\longrightarrow (F \wr G_1) \times (F \wr G_2) \\ (f, (g_1, g_2)) &\mapsto ((f|_{X_1}, g_1), (f|_{X_2}, g_2)) \end{aligned}$$

is a group isomorphism. $\qquad\qquad\square$

2.1.2 Composition and exponentiation actions

Now suppose in addition that F acts on a finite set Y. Then there is a natural action of $F \wr G$ on the product space $X \times Y$, as shown in the following.

Lemma 2.1.4   *For $(f, g) \in F \wr G$ and $(x, y) \in X \times Y$, set*

$$(f, g)(x, y) = (gx, f(gx)y) \equiv (gx, [(g^{-1}f)(x)]y). \tag{2.5}$$

Then (2.5) defines an action of $F \wr G$ on $X \times Y$. Moreover, (2.5) is transitive if and only if the actions of G on X and of F on Y are both transitive.

Proof   It is clear that $(1_F, 1_G)(x, y) = (x, y)$ for all $(x, y) \in X \times Y$. Moreover, if $(f, g), (f', g') \in F \wr G$ and $(x, y) \in X \times Y$ then

$$
\begin{aligned}
[(f, g)(f', g')](x, y) &= (f \cdot gf', gg')(x, y) \\
&= (gg'x, \{[(gg')^{-1}f \cdot g'^{-1}f'](x)\}y) \\
&= (gg'x, [(g^{-1}f)(g'x)]\{[(g'^{-1}f')(x)]y\}) \\
&= (f, g)(g'x, f'(g'x)y) \\
&= (f, g)[(f', g')(x, y)].
\end{aligned}
$$

It follows that (2.5) is an action. It is immediate to check that this action is transitive if and only if the actions of G on X and of F on Y are both transitive. $\qquad\square$

Definition 2.1.5   The action defined in (2.5) is called the *composition* of the actions of G on X and F on Y.

When restricted to the subgroup $(\operatorname{diag} F^X)G$, the composition action coincides with the product action of $G \times F$ on $X \times Y$. Note also that

$$(f, g)^{-1}(x, y) = (g^{-1}f^{-1}, g^{-1})(x, y) = (g^{-1}x, f(x)^{-1}y)$$

for all $(f, g) \in F \wr G$ and $x \in X, y \in Y$.

The theory of wreath products becomes more transparent if we think of them as groups acting on finite rooted trees. The *rooted tree* $T_{X \times Y}$ corresponding to $X \times Y$ is the graph (V, E) with vertex set $V = \{\emptyset\} \coprod X \coprod (X \times Y)$ and edge set $E = \{\{\emptyset, x\} : x \in X\} \coprod \{\{x, (x, y)\} : x \in X, y \in Y\}$. The vertex $\emptyset$ is called the *root* of $T_{X \times Y}$. Moreover, if $e = (u, v) \in E$, $u, v \in V$ then we say that the vertices u and v are adjacent and we write $u \sim v$. In other words, $T_{X \times Y}$ is the finite rooted tree, with two levels, obtained by taking $\emptyset$ as the root and X as the first level and then attaching to each $x \in X$ a copy of Y. All these copies of Y constitute the second level $X \times Y$, and its elements are called the *leaves* of $T_{X \times Y}$ (see Fig. 2.1).

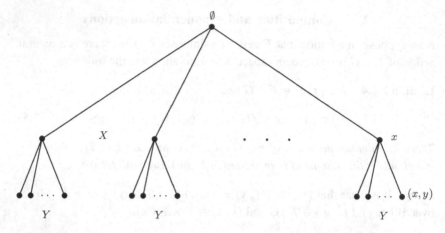

Fig. 2.1 The tree $T_{X \times Y}$ is obtained by attaching to each $x \in X$ a copy of Y.

We denote by $\text{Aut}(T_{X \times Y})$ the *automorphism group* of the tree $T_{X \times Y}$ of $T_{X \times Y}$, that is, the group of all bijective maps $\alpha \colon V \to V$ which respect the adjacency relation $\sim$ (in other words, which map edges to edges). Note that for every $\alpha \in \text{Aut}(T_{X \times Y})$ one has $\alpha(\emptyset) = \emptyset$ and, more generally, $\alpha(X) = X$ and $\alpha(X \times Y) = X \times Y$. In view of this, every $\alpha \in \text{Aut}(T_{X \times Y})$ is uniquely determined by its action on the leaves, that is, by $\alpha|_{X \times Y}$. Now, given $\alpha \in \text{Aut}(T_{X \times Y})$ and $(x, y) \in X \times Y$, there exist unique elements $x' = x'(x) \in X$ and $y' = y'(x, y) \in Y$ such that

$$\alpha(x, y) = (x', y'). \tag{2.6}$$

Taking $G = \text{Sym}(X)$ and $F = \text{Sym}(Y)$ in Lemma 2.1.4, we have that the composition action makes $\text{Sym}(Y) \wr \text{Sym}(X)$ act on the tree $T_{X \times Y}$. In fact we have the following.

Theorem 2.1.6 $\text{Aut}(T_{X \times Y}) \cong \text{Sym}(Y) \wr \text{Sym}(X)$.

Proof Note first that the composition action respects the adjacency relation $\sim$ on the vertices of $T_{X \times Y}$. Also, the unique element of $\text{Sym}(Y) \wr \text{Sym}(X)$ that corresponds to the trivial automorphism is the identity. This ensures that $\text{Sym}(Y) \wr \text{Sym}(X)$ can be identified with a subgroup of $\text{Aut}(T_{X \times Y})$. It only remains to show that every $\alpha \in \text{Aut}(T_{X \times Y})$ comes from the action of an element $(f, g) \in \mathrm{Sym}(Y) \wr \mathrm{Sym}(X)$. This is easy: define $g = g(\alpha) \in \mathrm{Sym}(X)$ and $f = f(\alpha) \in \mathrm{Sym}(Y)^X$ such that

$$gx = x' \qquad \text{and} \qquad f(gx)y = y'$$

where $\alpha(x, y) = (x', y')$, for all $x \in X$ and $y \in Y$. Note that f is well defined because for every $x \in X$ we have

$$\alpha(\{(x, y) : y \in Y\}) = \{(x', y') : y' \in Y\} = \{(gx, y) : y \in Y\}$$

(α respects the adjacency relation $\sim$). It is then clear that α and (f, g) yield the same action on $T_{X \times Y}$. $\square$

Returning to the composition action (see Definition 2.1.5), since $G \leq \mathrm{Sym}(X)$ and $F \leq \mathrm{Sym}(Y)$, from the above theorem we immediately deduce the following.

Corollary 2.1.7 (Geometric interpretation of the composition) *The group $F \wr G$ is isomorphic to a subgroup of* $\mathrm{Aut}(T_{X \times Y})$. $\square$

There is also a natural action of $F \wr G$ on the set Y^X of all maps $\varphi : X \to Y$.

Lemma 2.1.8 *For* $(f, g) \in F \wr G$, $\varphi \in Y^X$ *and* $x \in X$ *set*

$$[(f, g)\varphi](x) = f(x)\varphi(g^{-1}x). \tag{2.7}$$

Then (2.7) defines an action of $F \wr G$ on Y^X. Moreover, (2.7) is transitive if and only if the action of F on Y is transitive.

Proof It is clear that $(1_F, 1_G)\varphi = \varphi$. Moreover, if $(f, g), (f', g') \in F \wr G$, $\varphi \in Y^X$ and $x \in X$, we have

$$\begin{aligned}
\{[(f, g)(f', g')\varphi]\}(x) &= [(f \cdot gf', gg')\varphi](x) \\
&= [f(x)f'(g^{-1}x)]\varphi(g'^{-1}g^{-1}x) \\
&= f(x)[f'(g^{-1}x)\varphi(g'^{-1}g^{-1}x)] \\
&= f(x)\{[(f', g')\varphi](g^{-1}x)\} \\
&= \{(f, g)[(f', g')\varphi]\}(x)
\end{aligned}$$

and therefore (2.7) is an action. One can easily check that this action is transitive if and only if the action of F on Y is transitive (this is equivalent to a transitive action of the base group F^X itself on Y^X). $\square$

Definition 2.1.9 The action defined in (2.7) is called the *exponentiation* of the action of F by the action of G. Its restriction to (diag $F^X)G$ is called the *power action* of F by G.

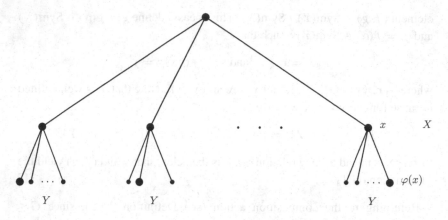

Fig. 2.2 The map $\varphi \in Y^X$ may be seen as a subtree of the tree of $X \times Y$.

Exercise 2.1.10   For each $\varphi \in Y^X$, denote by

$$\mathcal{G}_\varphi = \{(x, \varphi(x)) : x \in X\} \subset X \times Y$$

the *graph* of φ. Show that

$$(f, g)\mathcal{G}_\varphi = \mathcal{G}_{(f,g)\varphi},$$

where $(f, g)\mathcal{G}_\varphi = \{(f, g)(x, \varphi(x)) : x \in X\}$. In other words, on the family $\{\mathcal{G}_\varphi : \varphi \in Y^X\} \equiv Y^X$ of subsets of $X \times Y$ the composition action induces exactly the exponentiation action.

Example 2.1.11 (Geometric interpretation of the exponentiation action)
Consider again the tree $T_{X \times Y}$ of $X \times Y$. We may identify every $\varphi \in Y^X$ with the subtree

$$T_\varphi = \emptyset \bigsqcup X \bigsqcup \{(x, \varphi(x)) : x \in X)\}.$$

In this way, Y^X may be seen as the family of all subtrees whose vertex set consists of the root $\emptyset$, the whole first level X and, for every $x \in X$, the sole vertex $(x, \varphi(x)) \in X \times Y$ (see Fig. 2.2). Then the action of $F \wr G$ on this family of subtrees (induced by the composition action) coincides exactly with the exponentiation action (see Exercise 2.1.10).

2.1.3 Iterated wreath products and their actions on rooted trees

Now let H be a third group and suppose again that G (resp. F) acts transitively on X (resp. Y). We consider the wreath products $H \wr F \equiv H^Y \rtimes F$ and $(H \wr F) \wr G \equiv (H \wr F)^X \rtimes G$. Alternatively, if we regard $F \wr G$ as a group acting on $X \times Y$ by means of the composition action, we can form the wreath product $H \wr (F \wr G) \equiv H^{X \times Y} \rtimes (F \wr G)$. Both constructions lead to the same result, as shown in the following theorem.

Theorem 2.1.12 (Associativity of the wreath product)   *The map*

$$\Psi : \quad H \wr (F \wr G) \quad \to \quad (H \wr F) \wr G$$
$$(h, f, g) \quad \mapsto \quad (\vartheta, g),$$

where $h \in H^{X \times Y}$, $f \in F^X$ and $g \in G$ and where $\vartheta \in (H \wr F)^X = (H^Y \times F)^X = (H^Y)^X \times F^X$, defined by setting $\vartheta(x) = (h(x, \cdot), f(x))$ for all $x \in X$, is a group isomorphism.

Proof   It is clear that Ψ is a bijection. If we take $(h, f, g), (h', f', g') \in H \wr (F \wr G)$ then their product is

$$(h, f, g)(h', f', g') = (h \cdot (f, g)h', f \cdot gf', gg'),$$

where $[h \cdot (f, g)h'](x, y) = h(x, y)h'(g^{-1}x, f(x)^{-1}y)$. Therefore, on the one hand

$$\Psi((h, f, g)(h', f', g')) = (\vartheta'', gg'),$$

where $\vartheta''(x) = (h(x, \cdot)h'(g^{-1}x, f(x)^{-1}\cdot), f(x)f'(g^{-1}x))$. On the other hand, if $(\vartheta, g) = \Psi(h, f, g)$ and $(\vartheta', g') = \Psi(h', f', g')$ then $(\vartheta, g)(\vartheta', g') = (\vartheta \cdot g\vartheta', gg')$ and

$$(\vartheta \cdot g\vartheta')(x) = \vartheta(x)\vartheta'(g^{-1}x) = (h(x, \cdot), f(x))(h'(g^{-1}x, \cdot), f'(g^{-1}x))$$
$$= (h(x, \cdot)h'(g^{-1}x, f(x)^{-1}\cdot), f(x)f'(g^{-1}x)),$$

that is, $\vartheta \cdot g\vartheta' = \vartheta''$. This shows that

$$\Psi((h, f, g)(h', f', g')) = \Psi(h, f, g)\Psi(h', f', g'),$$

so that Ψ is an isomorphism. $\square$

From now on, in view of Theorem 2.1.12 we will simply write the iterated wreath product of H, F and G as $H \wr F \wr G$.

More generally, suppose that $G_1, G_2, \ldots, G_m$ are finite groups and that each G_i acts on a finite set X_i, $i = 1, 2 \ldots, m - 1$. Set $V_0 = \{\emptyset\}$ and, for $k = 1, 2, \ldots, m$,

$$V_k = X_1 \times X_2 \times \cdots \times X_k.$$

Then the *iterated wreath product* $G_m \wr G_{m-1} \wr \cdots \wr G_2 \wr G_1$ consists of all m-tuples $(f_m, f_{m-1}, \ldots, f_2, f_1)$, where $f_1 \in G_1$ and $f_k : V_{k-1} \to G_k, k = 2,$ $3, \ldots, m$, with the multiplication law and action on V_m recursively defined by

$$(f_k, f_{k-1}, \ldots, f_2, f_1)(f_k', f_{k-1}', \ldots, f_2', f_1')$$
$$= (f_k \cdot (f_{k-1}, \ldots, f_2, f_1)f_k', (f_{k-1}, \ldots, f_2, f_1)(f_{k-1}', \ldots, f_2', f_1')), \tag{2.8}$$

where

$$[(f_{k-1}, f_{k-2}, \ldots, f_2, f_1)f_k'](x_1, x_2, \ldots, x_{k-1})$$
$$= f_k'((f_{k-1}, \ldots, f_2, f_1)^{-1}(x_1, x_2, \ldots, x_{k-1})), \tag{2.9}$$

and by

$$(f_{k-1}, f_{k-2}, \ldots, f_2, f_1)(x_1, x_2, \ldots, x_{k-1}) = (y_1, y_2, \ldots, y_{k-1}), \tag{2.10}$$

where

$$(y_1, y_2, \ldots, y_{k-2}) = (f_{k-2}, \ldots, f_2, f_1)(x_1, x_2, \ldots, x_{k-2})$$

and

$$y_{k-1} = f_{k-1}(y_1, y_2, \ldots, y_{k-2})x_{k-1}$$

for all $x_i \in X_i$, $f_i \in G_i^{V_{i-1}}$, $i = 1, 2, \ldots, k$, and $k = 1, 2 \ldots, m$.

Exercise 2.1.13   Verify that the multiplication operation defined in (2.8) makes $G_m \wr G_{m-1} \wr \cdots \wr G_2 \wr G_1$ a group and that (2.10) defines a group action of $G_m \wr G_{m-1} \wr \cdots \wr G_2 \wr G_1$ on $X_1 \times X_2 \times \cdots \times X_m$.
Hint. Apply induction and use Lemmas 2.1.1 and 2.1.4.

Exercise 2.1.14 (Distributivity of the wreath product)   Let G_1 (resp. G_2) be a finite group acting on a set X_1 (resp. X_2). Let F be a finite group and set $X = X_1 \coprod X_2$. Show that

$$F \wr (G_1 \times G_2) \cong (F \wr G_1) \times (F \wr G_2),$$

where the wreath product on the left-hand side is defined with respect to the action of $G_1 \times G_2$ on X (see Proposition 2.1.3).
Hint. Show that the map

$$\begin{aligned} F^X \rtimes (G_1 \times G_2) &\to (F^{X_1} \rtimes G_1) \times (F^{X_2} \rtimes G_2) \\ (f, (g_1, g_2)) &\mapsto ((f|_{X_1}, g_1), (f|_{X_2}, g_2)), \end{aligned}$$

where $f|_{X_i} \in F^{X_i}$ denotes the restriction of $f \in F^X$, yields the desired isomorphism.

The iterated wreath product of permutation representations is a particular case of a more general construction considered in [3]. It involves an indicized family of permutation groups where the index set is a poset.

2.1.4 Spherically homogeneous rooted trees and their automorphism group

In this subsection, we give a geometric interpretation of the iterated wreath product: as for the wreath product of two single groups, the iterated wreath product can be interpreted in terms of actions by automorphisms of rooted trees.

Let $X_1, X_2, \ldots, X_m, m \geq 1$, be finite sets. For $k = 1, 2, \ldots, m$, let $r_k = |X_k|$ and

$$V_k = X_1 \times X_2 \times \cdots \times X_k$$

and set $\mathbf{r} = (r_1, r_2, \ldots, r_m)$. We then denote by $T_{\mathbf{r}}$ the *spherically homogeneous rooted tree of branching type* $\mathbf{r}$ (briefly, $\mathbf{r}$-*tree*) with vertex set

$$V = V_0 \coprod V_1 \coprod V_2 \coprod \cdots \coprod V_m,$$

where two vertices $v = (x_1, x_2, \ldots, x_k)$ and $w = (y_1, y_2, \ldots, y_h)$ are adjacent if $|h - k| = 1$ and $x_i = y_i$ for all $i = 1, 2, \ldots, \min\{h, k\}$. If v and w are adjacent and $h = k + 1$ we say that w is a *son* or *successor* of v and that v is a *father* or *predecessor* of w. The set $V_k, k = 0, 1, \ldots, m$, is called the *kth level* of the tree. Clearly, every vertex of level k ($0 \leq k \leq m - 1$) has exactly r_{k+1} successors. Moreover, the integer m is called the *depth* of $T_{\mathbf{r}}$ and the elements in V_m are called the *leaves* of $T_{\mathbf{r}}$. Let $\mathrm{Aut}(T_{\mathbf{r}})$ denote the group of all rooted automorphisms of $T_{\mathbf{r}}$, that is, of all bijective maps $\alpha \colon V \to V$ that preserve the adjacency relation and that, in addition, fix the root $\emptyset$. Note that every such automorphism stabilizes all levels of the tree. For every vertex $(x_1, x_2, \ldots, x_k) \in V_k, 0 \leq k \leq m$, we denote by $T_{(x_1, x_2, \ldots, x_k)}$ the subtree with root $(x_1, x_2, \ldots, x_k)$ and vertex set

$$\{(x_1, x_2, \ldots, x_k)\} \coprod (\{(x_1, x_2, \ldots, x_k)\} \times X_{k+1}) \coprod$$
$$\cdots \coprod (\{(x_1, x_2, \ldots, x_k)\} \times X_{k+1} \times X_{k+2} \times \cdots \times X_m).$$

The following constitutes a generalization of Theorem 2.1.6.

Theorem 2.1.15 $\mathrm{Aut}(T_{\mathbf{r}}) \cong S_{r_m} \wr S_{r_{m-1}} \wr \cdots \wr S_{r_2} \wr S_{r_1}$.

Proof   Arguing by induction, one easily proves that the iterated composition action respects the adjacency relation $\sim$ on the vertices of $T_{\mathbf{r}}$ and that the unique element of $S_{r_m} \wr S_{r_{m-1}} \wr \cdots \wr S_{r_2} \wr S_{r_1}$ which corresponds to the trivial automorphism is the identity. This ensures that $S_{r_m} \wr S_{r_{m-1}} \wr \cdots \wr S_{r_2} \wr S_{r_1}$ can be identified with a subgroup of $\text{Aut}(T_r)$. It remains to show that every $\alpha \in \text{Aut}(T_r)$ comes from the action of an element in the iterated wreath product. The argument is the same as that in Theorem 2.1.6, just a bit more elaborated.

If $\alpha \in \text{Aut}(T_r)$ and $\alpha(x_1, x_2, \ldots, x_k) = (y_1, y_2, \ldots, y_k)$ then, since α preserves the adjacency relation as well as the levels, we have $\alpha T_{(x_1, x_2, \ldots, x_k)} = T_{(y_1, y_2, \ldots, y_k)}$. In other words, every automorphism permutes the subtrees rooted at vertices of the same level. In this way, every $\alpha \in \text{Aut}(T_r)$ is uniquely determined by the m-tuple $h(\alpha) = (h_m, h_{m-1}, \ldots, h_2, h_1)$, where

$$h_k = h_k(\alpha) \in \text{Sym}(X_k)^{V_{k-1}}$$

is defined by

$$\alpha(x_1, x_2, \ldots, x_k)$$
$$= (h_1(\emptyset)x_1, h_2(x_1)x_2, h_3(x_1, x_2)x_3, \ldots, h_k(x_1, x_2, \ldots, x_{k-1})x_k) \quad (2.11)$$

for all $k = 1, 2, \ldots, m$. One calls $h(\alpha)$ the *labeling* (or *portrait*, see [33]) of the automorphism α.

We now recursively define $f_k \in \text{Sym}(X_k)^{V_{k-1}}$ by setting $f_1 = h_1$ and

$$f_k = (f_{k-1}, f_{k-2}, \ldots, f_2, f_1)h_k$$

for all $k = 2, 3, \ldots, m$ and denote by

$$f(\alpha) = (f_m, f_{m-1}, \ldots, f_2, f_1)$$

the corresponding m-tuple.

Then (2.11) ensures that α can be identified with $f(\alpha)$. $\square$

2.1.5 The finite ultrametric space

The finite ultrametric space has been introduced and studied by G. Letac [52] and A. Figà-Talamanca [28]. We refer to [11, Chapter 7] for more details and information.

Let q and m be positive integers and denote by $T_{q,m}$ the rooted tree T_r of depth m where $\mathbf{r} = (q, q, \ldots, q)$ (thus $T_{q,m}$ is homogeneous). Note that, as a particular case of Theorem 2.1.15, we have $\text{Aut}(T_{q,m}) \cong S_q \wr S_q \wr \cdots \wr S_q$.

Denote by $Y = \{0, 1, \ldots, q-1\}^m$ the set of leaves of $T_{q,m}$ and equip it with a metric structure by defining a distance function $d: Y \times Y \to \mathbb{N}$, by setting

$$d(x, y) = m - \max\{k : x_i = y_i \text{ for all } i \leq k\} \quad (2.12)$$

for all $x = (x_1, x_2, \ldots, x_m)$ and $y = (y_1, y_2, \ldots, y_m)$ in Y. The distance d satisfies the *ultrametric inequality*

$$d(x, z) \leq \max\{d(x, y), d(y, z)\}$$

for all $x, y, z \in Y$. For this reason, one calls (Y, d) the (finite) *ultrametric space*.

Consider the action of $\mathrm{Aut}(T_{q,m})$ on Y and denote by $K(q, m)$ the stabilizer of the point $y_0 = (0, 0, \ldots, 0)$. We then have:

Theorem 2.1.16 $(\mathrm{Aut}(T_{q,m}), K(q, m))$ *is a symmetric Gelfand pair.*

Proof   By virtue of Example 1.2.32, it suffices to show that the action of $\mathrm{Aut}(T_{q,m})$ on Y is 2-point homogeneous.

We proceed by induction on the depth m of the tree. If $m = 1$, we have $Y = \{0, 1, \ldots, q - 1\}$ and $\mathrm{Aut}(T_{q,1}) = S_q$. Moreover, the ultrametric distance coincides, in this case, with the discrete distance (for $x, y \in Y$, we have $d(x, y) = 0$ if $x = y$ and $d(x, y) = 1$ otherwise). Let $x, y, x', y' \in Y$ and suppose that $d(x, y) = d(x', y')$. We distinguish three cases: (i) if $d(x, y) = d(x', y') = 0$ then $x = y$ and $x' = y'$, and we denote by g the transposition $(x\ x') \in S_q$; (ii) if $d(x, y) = d(x', y') = 1$ (so that $x \neq y$ and $x' \neq y'$), $x = y'$ and $y = x'$, we then set $g = (x\ x') = (x\ y) \in S_q$; (iii) otherwise, we set $g = (x\ x')(y\ y') \in S_q$. In all three cases we have $gx = x'$ and $gy = y'$. This proves the base case of the induction.

Suppose now that the statement holds true for $\mathrm{Aut}(T_{q,k})$ with $1 \leq k \leq m-1$. Let $x = (x_1, x_2, \ldots, x_m)$, $y = (y_1, y_2, \ldots, y_m)$, $x' = (x'_1, x'_2, \ldots, x'_m)$ and $y' = (y'_1, y'_2, \ldots, y'_m)$.

If $d(x, y) = d(x', y') = m$ then $x_1 \neq y_1$ and $x'_1 \neq y'_1$. By applying the same argument as in the base case, we can find an element $g \in S_q$ such that $gx_1 = x'_1$ and $gy_1 = y'_1$. Consider now the element $\alpha \in \mathrm{Aut}(T_{q,m})$ with label $h(\alpha) = (1, 1, \ldots, 1, g)$ and set $x'' = (x''_1, x''_2, \ldots, x''_m) = \alpha(x) = (x'_1, x_2, \ldots, x_m)$ and $y'' = (y''_1, y''_2, \ldots, y''_m) = \alpha(y) = (y'_1, y_2, \ldots, y_m)$. We have that x'' and x' (resp. y'' and y') belong to the same rooted subtree, namely $T_{x'_1}$ (resp. $T_{y'_1}$), and these subtrees are distinct (since $x'_1 \neq y'_1$). Since the height of these rooted trees is $m-1$, after identifying $\mathrm{Aut}(T_{x'_1})$ and $\mathrm{Aut}(T_{y'_1})$ with $\mathrm{Aut}(T_{q,m-1})$ we apply induction and obtain an element $\beta \in \mathrm{Aut}(T_{q,m-1})$ (resp. $\delta \in \mathrm{Aut}(T_{q,m-1})$), say with label $h(\beta) = (b_m, b_{m-1}, \ldots, b_2)$ (resp. $h(\delta) = (d_m, d_{m-1}, \ldots, d_2)$), such that $\beta(x''_2, x''_3, \ldots, x''_m) = (x'_2, x'_3, \ldots, x'_m)$ (resp. $\delta(y''_2, y''_3, \ldots, y''_m) = (y'_2, y'_3, \ldots, y'_m)$). If $\gamma \in \mathrm{Aut}(T_{q,m})$ has a label

$h(\gamma) = (h_m, h_{m-1}, \ldots, h_2, h_1)$ such that $h_1 = g$, $h_k(x_1', t_2, t_3, \ldots, t_{k-1}) = b_k(t_2, t_3, \ldots, t_{k-1})$ and $h_k(y_1', t_2, t_3, \ldots, t_{k-1}) = d_k(t_2, t_3, \ldots, t_{k-1})$ for all $t_2, t_3, \ldots, t_{k-1} = 0, 1, \ldots, q - 1$ and $2 \leq k \leq m$ then we clearly have $\gamma(x) = x'$ and $\gamma(y) = y'$.

Suppose now that $d(x, y) = d(x', y') < m$. This means that there exists $1 \leq n \leq m$ such that $x_1 = y_1, x_2 = y_2, \ldots, x_n = y_n$ and $x_1' = y_1', x_2' = y_2', \ldots, x_n' = y_n'$. In particular $x_1 = y_1$ (resp. $x_1' = y_1'$), that is, x and y (resp. x' and y') belong to the same subtree T_{x_1} (resp. $T_{x_1'}$). By the transitivity of the action of S_q on $\{0, 1, \ldots, q - 1\}$, there exists $g \in S_q$ such that $gx_1 = x_1'$ (and therefore $gy_1 = y_1'$). Let $\alpha \in \mathrm{Aut}(T_{q,m})$ denote the element with label $h(\alpha) = (1, 1, \ldots, 1, g)$ and set $x'' = (x_1'', x_2'', \ldots, x_m'') = \alpha(x) = (x_1', x_2, \ldots, x_m)$ and $y'' = (y_1'', y_2'', \ldots, y_m'') = \alpha(y) = (x_1', y_2, \ldots, y_m)$. Then, after identifying $\mathrm{Aut}(T_{x_1'})$ and $\mathrm{Aut}(T_{q,m-1})$, we can find, by the inductive hypothesis, an element $\beta \in \mathrm{Aut}(T_{q,m-1})$ such that $\beta(x_2'', \ldots, x_m'') = (x_2', \ldots, x_m')$ and $\beta(y_2'', \ldots, y_m'') = (y_2', \ldots, y_m')$. Finally, let $\gamma \in \mathrm{Aut}(T_{q,m})$ be an element with label $h(\gamma) = (h_m, h_{m-1}, \ldots, h_2, h_1)$ such that $h_1 = g$ and $h_k(x_1', t_2, t_3, \ldots, t_{k-1}) = b_k(t_2, t_3, \ldots, t_{k-1})$ for all $t_2, t_3, \ldots, t_{k-1} = 0, 1, \ldots, q - 1$ and $2 \leq k \leq m$, where $h(\beta) = (b_m, b_{m-1}, \ldots, b_2)$. It is clear that $\gamma(x) = x'$ and $\gamma(y) = y'$. $\square$

In order to describe the decomposition into irreducible subrepresentations of the permutation representation $L(Y)$, we introduce some subspaces. Recalling that $Y = \{0, 1, \ldots, q - 1\}^m$, we regard any element $f \in L(Y)$ as a function $f = f(x_1, x_2, \ldots, x_m)$ of the variables $x_1, x_2, \ldots, x_m$ ranging in $\{0, 1, \ldots, q - 1\}$. Moreover, if $0 \leq j \leq m - 1$ and if $f \in L(Y)$ does not depend on the last variables $x_{j+1}, x_{j+1}, \ldots, x_m$, that is, f only depends on the vertices of the first j levels of the tree $T_{q,m}$, then we simply write $f = f(x_1, x_2, \ldots, x_j)$.

Set $W_0 = L(\emptyset) = \mathbb{C}$ and, for $j = 1, \ldots, m$, set

$$W_j = \left\{ f \in L(Y) : f = f(x_1, x_2, \ldots, x_j), \sum_{x=0}^{q-1} f(x_1, x_2, \ldots, x_{j-1}, x) \equiv 0 \right\}.$$

$$(2.13)$$

Note that for $j \geq 1$ one has

$$\dim(W_j) = q^{j-1}(q - 1). \tag{2.14}$$

In other words, W_j is the set of all functions $f \in L(Y)$ that depend only on the variables $x_1 x_2 \cdots x_j$ and whose mean on the sets $\{x_1 x_2 \cdots x_{j-1} x : x = 0, 1, \ldots, q - 1\}$ is equal to zero for all $x_1 x_2 \cdots x_{j-1} \in \{0, 1, \ldots, q - 1\}^{j-1}$.

In more geometrical language, we may say that $f \in L(Y)$ belongs to W_j if, for every $x_1 x_2 \cdots x_{j-1} \in \{0, 1, \ldots, q-1\}^{j-1}$ and $x \in \{0, 1, \ldots, q-1\}$, the function f is constant on the set

$$A_x = \{x_1 x_2 \cdots x_{j-1} x x_{j+1} x_{j+2} \cdots x_\ell :$$
$$x_{j+1}, x_{j+2}, \ldots, x_\ell = 0, 1, \ldots, q-1 \text{ and } \ell = j+1, j+2, \ldots, m\}$$

of descendants of $x_1 x_2 \cdots x_{j-1} x$ in Y and, denoting by f_x the constant value of f on A_x, one has $\sum_{x=0}^{q-1} f_x = 0$.

Theorem 2.1.17 $L(Y) = \oplus_{j=0}^{m} W_j$.

Proof We first show that the subspaces W_j, $j = 0, 1, \ldots, m$, are $\mathrm{Aut}(\mathcal{T}_{q,m})$-invariant. The first condition, namely the dependence of $f \in W_j$ only on the first j variables, is clearly invariant as $\mathrm{Aut}(\mathcal{T}_{q,m})$ preserves the levels.

The induced action of $\mathrm{Aut}(\mathcal{T}_{q,m})$ on $L(Y)$ is given by

$$[\alpha^{-1} f](x_1, x_2, \ldots, x_n)$$
$$= f(h(\emptyset)x_1, h(x_1)x_2, h(x_1, x_2)x_3, \ldots, h(x_1, x_2, \ldots, x_{m-1})x_m)$$

for $f \in L(Y)$ and $\alpha \in \mathrm{Aut}(\mathcal{T}_{q,m})$ with labeling as in (2.11). Thus if $f \in W_j$ we have

$$\sum_{x=0}^{q-1} [\alpha^{-1} f](x_1, x_2, \ldots, x_{j-1}, x)$$

$$= \sum_{x=0}^{q-1} f(h(\emptyset)x_1, h(x_1)x_2,$$

$$\ldots, h(x_1, x_2, \ldots, x_{j-2})x_{j-1}, h(x_1, x_2, \ldots, x_{j-1})x)$$

$$= \sum_{x'=0}^{q-1} f(h(\emptyset)x_1, h(x_1)x_2, \ldots, h(x_1, x_2, \ldots, x_{j-2})x_{j-1}, x')$$

$$\equiv 0.$$

This shows that the second defining condition for an f to be in W_j is also invariant.

We now show that these spaces are pairwise orthogonal; that is, if $f \in W_j$, $f' \in W_{j'}$ then $\langle f, f' \rangle = 0$ if $j \neq j'$. To fix our ideas suppose that $j < j'$. Then

$$\langle f, f' \rangle = \sum_{x_1=0}^{q-1} \sum_{x_2=0}^{q-1} \cdots \sum_{x_m=0}^{q-1} f(x_1, x_2, \ldots, x_m) \overline{f'(x_1, x_2, \ldots, x_m)}$$

$$= q^{m-j'} \sum_{x_1=0}^{q-1} \sum_{x_2=0}^{q-1} \cdots \sum_{x_{j'-1}=0}^{q-1} f(x_1, x_2, \ldots, x_j)$$

$$\times \sum_{k=0}^{q-1} \overline{f'(x_1, x_2, \ldots, x_{j'-1}, k)}$$

$$= 0.$$

We claim that the W_j's fill up the whole space $L(Y)$. We use induction on m. For $m = 1$, it is a standard fact that any function $f(x_1)$ can be expressed as $f(x_1) = c + g(x_1)$, where $c \in \mathbb{C}$ is a constant (indeed, $c = \frac{1}{q} \sum_{x=0}^{q-1} f(x)$) and g is a function of mean zero: $\sum_{x=0}^{q-1} g(x) = 0$. Suppose now that the assertion is true for $n - 1$. Again, we can express an element $f \in L(Y)$ as $f(x_1, x_2, \ldots, x_m) = c(x_1, x_2, \ldots, x_{m-1}) + g(x_1, x_2, \ldots, x_{m-1}, x_m)$, where c does not depend on the last variable x_m and g has mean zero with respect to x_m. Applying the inductive step to c, the claim follows.

For $j = 0, 1, \ldots, m$ we denote by $\Omega_j = \{x \in Y : d(x, x_0) = j\}$ the sphere of radius j centered at $y_0 = (0, 0, \ldots, 0)$. Observe that since $K(q, n)$ is the stabilizer of the point y_0 the Ω_j's are $K_{q,m}$-invariant. In fact, by virtue of the 2-point homogeneity of the action of $\mathrm{Aut}(T_{q,m})$ on Y, the spheres Ω_j are exactly the $K_{q,m}$-orbits. It follows that there are $m + 1$ such orbits.

By applying Theorem 1.2.36 we have that the W_j's are irreducible subspaces, and this ends the proof. $\qquad\qquad\square$

We remark that, incidentally, Theorem 2.1.17 offers an alternative proof of the fact that $(\mathrm{Aut}(T_{q,n}), K(q, n))$ is a Gelfand pair (compare with Theorem 2.1.16). Indeed the subspaces W_j are pairwise inequivalent (they have different dimensions, see (2.14)) and therefore the decomposition of $L(Y)$ is multiplicity free (cf. Theorem 1.2.28).

Our next step is the determination of the spherical functions $\phi_0, \phi_1, \ldots, \phi_m$ relative to $(\mathrm{Aut}(T_{q,m}), K(q, m))$: we combine the defining conditions of the W_j's with $K(q, m)$-invariance (ϕ_j is constant on the spheres Ω_k).

Proposition 2.1.18   *For $j = 0, 1, \ldots, m$, the spherical function $\phi_j \in W_j$ is given by*

$$
\phi_j(x) = \begin{cases} 1 & \text{if } d(x, x_0) < m - j + 1 \\ -\frac{1}{q-1} & \text{if } d(x, x_0) = m - j + 1 \\ 0 & \text{if } d(x, x_0) > m - j + 1. \end{cases} \tag{2.15}
$$

Proof   It is clear that the function in (2.15) is $K(q, m)$-invariant. We are only left with showing that ϕ_j belongs to W_j. By virtue of (2.12) we have

$$
\phi_j(x_1, x_2, \ldots, x_m) = \begin{cases} 1 & \text{if } x_1 = x_2 = \cdots = x_j = 0 \\ -\frac{1}{q-1} & \text{if } x_1 = x_2 = \cdots = x_{j-1} = 0 \text{ and } x_j \neq 0 \\ 0 & \text{otherwise.} \end{cases}
$$

Indeed, we first observe that if $(x_1, x_2, \ldots, x_{j-1}) \neq (0, 0, \ldots, 0)$ then all points of the form $(x_1, x_2, \ldots, x_m)$ have the same distance ($> m - j + 1$) from the base point $y_0 = (0, 0, \ldots, 0)$ (in fact all points in the spheres Ω_h with $h > m - j + 1$ satisfy this condition).

The spherical function ϕ_j in W_j is constant on the Ω_h's and thus if $(x_1, x_2, \ldots, x_{j-1}) \neq (0, 0, \ldots, 0)$ then $\phi_j(x_1, x_2, \cdots, x_{j-1}, x)$ does not depend on x because all the $(x_1, x_2, \ldots, x_{j-1}, x)$, with $x = 0, 1, \ldots, q - 1$, belong to the same orbit Ω_h. This, coupled with the condition

$$
\sum_{x=0}^{q-1} \phi_j(x_1, x_2, \ldots, x_{j-1}, x) = 0,
$$

implies that ϕ_j vanishes on all points at distance $> m - j + 1$ from x_0.

Similarly, the points of the form $(\underbrace{0, 0, \ldots, 0}_{j-1}, x, y_{j+1}, \ldots, y_m)$ with $x = 1, 2, \ldots q - 1$ constitute the ball of radius $m - j + 1$. Since, by definition, $\phi_j(0, 0, \ldots, 0, 0) = 1$ and ϕ_j only depends on the first j variables, the condition $\phi_j(0, 0, \ldots, 0, 0) + \sum_{x=1}^{q-1} \phi_j(0, 0, \ldots, 0, x) = 0$, coupled with the condition that ϕ_j is constant on Ω_{m-j+1}, uniquely determines the value of ϕ_j on points at distance $m - j + 1$; this value is therefore equal to $-\frac{1}{q-1}$.

Finally, if $d(x, x_0) < m - j + 1$ then $x = (\underbrace{0, 0, \ldots, 0}_{h}, y_{h+1}, \ldots, y_n)$ with $h > j - 1$, and therefore $\phi_j(x) = \phi_j(0, 0, \ldots, 0) = 1$. $\square$

2.2 Two applications of wreath products to group theory

A fundamental application of wreath products to group theory consists in expressing the Sylow p-subgroups $\mathrm{Sylow}(S_{p^n})$ of the symmetric group S_{p^n} of degree p^n (p a prime number, $n \geq 1$) as the n-iterated wreath product of the cyclic group C_p of order p; that is, we write

$$\mathrm{Sylow}(S_{p^n}) \cong \underbrace{C_p \wr C_p \wr \cdots \wr C_p}_{n \text{ times}}. \tag{2.16}$$

In general, if $N = a_0 + a_1 p + \cdots + a_k p^k$, where $0 \leq a_i < p$, then the Sylow p-subgroups of the symmetric group of degree N are isomorphic to the direct product of a_i copies of $\mathrm{Sylow}(S_{p^i})$ for $i = 1, 2, \ldots, k$.

These calculations are attributed to Kaloujnine [39–42] (see also [8]) although Kerber [43, p. 26] refers to an 1844 work of Cauchy. We shall not discuss these results here; the interested reader may find a detailed exposition in the book by Rotman [60, Chapter VII, p. 176].

In this section, however, we give two other applications of wreath products to group theory. They are relevant to the material in the present book. We follow quite closely the monograph by Dixon and Mortimer [23].

2.2.1 The theorem of Kaloujnine and Krasner

Let G, K and N be groups. We say that G is an *extension* of N by K, or equivalently that

$$\{1\} \longrightarrow N \longrightarrow G \longrightarrow K \longrightarrow \{1\}$$

is an *exact sequence*, if

$$N \trianglelefteq G \quad \text{and} \quad G/N \cong K.$$

A wreath product of the form $N \wr K = N^K \rtimes K$ (here we consider the left regular action of K on itself) yields an extension of the base group N^K by K. We refer to this as to the *regular* (or *standard*) wreath product of N by K.

Theorem 2.2.1 (Kaloujnine–Krasner [46–48]) *Let G be an extension of N by K. Then the following hold.*

(i) *There exists an injective homomorphism*

$$\Phi : G \longrightarrow N \wr K$$

of G into the regular wreath product of N by K.

(ii) *Moreover, if N^K is the base group of $N \wr K$ then*

$$\Phi(N) = \Phi(G) \cap N^K.$$

Proof (i) Let $\psi : G \longrightarrow K$ denote the canonical surjective homomorphism with Ker $\psi = N$. For every $k \in K$ choose $s_k \in G$ such that $\psi(s_k) = k$ and $s_{1_K} = 1_G$. In this way we have

$$G = \coprod_{k \in K} s_k N.$$

Since $\psi(s_k^{-1} g s_{\psi(g)^{-1}k}) = k^{-1} \psi(g) \psi(g)^{-1} k = 1_K$, then $s_k^{-1} g s_{\psi(g)^{-1}k} \in N$ for all $g \in G$ and $k \in K$. Therefore for every $g \in G$ we have a map $f_g : K \longrightarrow N$ defined by setting

$$f_g(k) = s_k^{-1} g s_{\psi(g)^{-1}k}, \tag{2.17}$$

for all $k \in K$. We claim that the map

$$\begin{array}{rcl} \Phi : & G & \longrightarrow & N \wr K \\ & g & \longmapsto & (f_g, \psi(g)) \end{array} \tag{2.18}$$

is an injective homomorphism. First, for $g, h \in G$ and $k \in K$ we have

$$\left\{ f_g [\psi(g) f_h] \right\} (k) = f_g(k) f_h \left[\psi(g)^{-1} k \right] \qquad \text{(by (2.2))}$$

$$= s_k^{-1} g s_{\psi(g)^{-1}k} s_{\psi(g)^{-1}k}^{-1} h s_{\psi(h)^{-1}\psi(g)^{-1}k}^{-1} \qquad \text{(by (2.17))}$$

$$= s_k^{-1} g h s_{\psi((gh)^{-1})k} \qquad (\psi \text{ is a homomorphism})$$

$$= f_{gh}(k) \qquad \text{(again by (2.17))},$$

that is,

$$f_g[\psi(g) f_h] = f_{gh} \qquad \text{for all } g, f \in G. \tag{2.19}$$

Then

$$\Phi(g)\Phi(h) = (f_g, \psi(g))(f_h, \psi(h)) \qquad \text{(by (2.18))}$$

$$= (f_g \psi(g) f_h, \psi(g)\psi(h)) \qquad \text{(by (2.1))}$$

$$= (f_{gh}, \psi(gh)) \qquad \text{(by (2.19))}$$

$$= \Phi(gh) \qquad \text{(by (2.18))}$$

for all $g, h \in G$, showing that Φ is a homomorphism. It remains to show that Φ is injective, that is, that Ker $\Phi = 1_G$. This is easy: if $\Phi(g) = 1_{N \wr K}$ then $f_g = 1_N$ and $\psi(g) = 1_K$ and therefore

$$1_G \equiv 1_N = 1_N(1_K) = f_g(1_K) = s_{1_K}^{-1} g s_{\psi(g)^{-1} 1_K} = g.$$

In order to show (ii), it suffices to note that $\Phi(g) \in N^K$ if and only if $\psi(g) = 1_K$, that is, if and only if $g \in N$. $\qquad\qquad\qquad\qquad\square$

2.2.2 Primitivity of the exponentiation action

Let G be a finite group acting transitively on a finite set X and denote by K the stabilizer of a point $x_0 \in X$.

A *block* of the action of G on X is a subset $B \subseteq X$ such that for each $g \in G$ one has

$$\text{either} \quad gB = B \quad \text{or} \quad gB \cap B = \emptyset. \tag{2.20}$$

Let $B \subset X$ be a block. We say that B is *trivial* if $|B| = 1$ or $B = X$. Setting $\Omega = \{kB : k \in K\}$, we have that $X = \coprod_{A \in \Omega} A$ and that G acts on Ω in the obvious way.

If there exists a *nontrivial* block, say B, we say that the action of G on X is *imprimitive* and we call Ω a *system of blocks* (or *system of imprimitivity*) for the action of G on X. If the action of G on X (is transitive and) has only trivial blocks we say that it is *primitive*.

Lemma 2.2.2   *Let $B \subset X$ be a block and $h \in G$. Then hB is also a block. Moreover, B is trivial if and only if hB is.*

Proof   The statement follows immediately after taking $h^{-1}gh$ in place of g in (2.20). The remaining part of the proof is trivial. $\qquad\qquad\qquad\square$

Exercise 2.2.3   Prove that the action of G on X is primitive if and only if K is a maximal subgroup of G.
Hint. The map $H \longmapsto Hx_0$ yields a correspondence between the set of all subgroups H such that $K \leq H$ and the set of all blocks containing x_0.

We recall that, in the preceding notation, the stabilizer of an element $x \in X$ is gKg^{-1}, where g is any group element such that $gx_0 = x$. A transitive action of G on X is called *regular* whenever the stabilizer of a point is trivial: $K = \{1_G\}$ (and therefore all other stabilizers are trivial). Clearly, the action of G is regular on X if and only if the following condition is satisfied: for any $x, y \in X$ there exists exactly one $g \in G$ such that $gx = y$. In particular, $|G| = |X|$.

The *kernel* of the action of G on X is the normal subgroup $N = \{g \in G : gx = x$ for all $x \in X\}$. This is the kernel of the homomorphism $G \to \text{Sym}(X)$ induced by the action. Note that the kernel coincides with K if and only if K is normal in G. We say that the action is *faithful* when its kernel is trivial.

Exercise 2.2.4 Suppose that the action of G on X is faithful and primitive. Then the action is not regular if and only if $N_G(K) = K$, where $N_G(K) = \{g \in G : gKg^{-1} = K\} \le G$ is the *normalizer* of K in G.
Hint. Observe that $K \le N_G(K) \trianglelefteq G$.

Denoting by N the kernel of the action of G on X, we have that the action is primitive if and only if the induced action of G/N on X is primitive. Therefore, in the study of primitive actions, the assumption that these actions are faithful is not restrictive.

Now let F be another group acting on a finite set Y.

On the one hand, the composition action of wreath products is not primitive because, in the notation of Definition 2.1.5, we have that $\{x\} \times Y$ is a block (for the action of $F \wr G$ on $X \times Y$) for every $x \in X$.

On the other hand, we will show that exponentiation (see Definition 2.1.9) provides a wealth of primitive actions. We first introduce some notation. Fix $y_0 \in Y$ and let H be the stabilizer of y_0 in F. For each $x \in X$ we set $\Psi_x = \{\psi \in Y^X : \psi(x') = y_0 \text{ for all } x' \ne x\}$.

Lemma 2.2.5 *Let Φ be a block for the action of $F \wr G$ on Y^X. Then, for every $x \in X$, the set $B = \{\varphi(x) : \varphi \in \Psi_x \cap \Phi\}$ is a block for the action of F on Y.*

Proof Let $x \in X$, $\varphi \in \Psi_x$ and suppose that $f \in F^X$ satisfies $f(x') = 1_F$ for all $x' \ne x$. Then one has

$$[(f, 1_G)\varphi](x) = f(x)\varphi(x) \quad \text{and} \quad [(f, 1_G)\varphi](x') = y_0 \quad \text{for all } x' \ne x,$$

and therefore $(f, 1_G)\Psi_x = \Psi_x$ and $(f, 1_G)(\Psi_x \cap \Phi) = \Psi_x \cap (f, 1_G)\Phi$. Since Φ is a block, we have

$$(f, 1_G)(\Psi_x \cap \Phi) = \Psi_x \cap \Phi \quad \text{or} \quad [(f, 1_G)(\Psi_x \cap \Phi)] \cap (\Psi_x \cap \Phi) = \emptyset.$$

Suppose now that $s \in F$ and $f \in F^X$ satisfies $f(x) = s$ and $f(x') = 1_F$ for all $x' \ne x$. Then we have

$$sB = \{\varphi(x) : \varphi \in (f, 1_G)(\Psi_x \cap \Phi)\}$$

so that, necessarily, either $sB = B$ or $sB \cap B = \emptyset$, proving that B is a block. □

Theorem 2.2.6 *Suppose that the actions of G on X and of F on Y are both faithful. Then the exponentiation action of $F \wr G$ on Y^X is primitive if and only if the following conditions are satisfied:*

(i) *the action of G on X is transitive;*
(ii) *the action of F on Y is primitive but not regular.*

Proof We start by proving that conditions (i) and (ii) are necessary. We achieve this by analyzing all possible cases separately.

- If the action of G on X is not transitive and $A \subsetneq X$, $A \neq \emptyset$, is an orbit, then the set $\Phi = \{\varphi \in Y^X : \varphi(x) = y_0 \text{ for all } x \in A\}$ is a nontrivial block of $F \wr G$ on Y^X. Indeed, if $(f, g) \in F \wr G$ and $(f, g)\Phi \cap \Phi \neq \emptyset$ then $f(x) \in H$ for all $x \in A$ and therefore $(f, g)\Phi = \Phi$.

- If the action of F on Y is not transitive then neither is the action of $F \wr G$ on Y^X transitive (see Lemma 2.1.8).

- If the action of F on Y is not primitive and $B \subseteq Y$ is a nontrivial block then the set $\Phi = \{\varphi \in Y^X : \varphi(x) \in B \text{ for all } x \in X\}$ is a nontrivial block for the action of $F \wr G$ on Y^X.

- If the action of F on Y is regular then the set Φ of all constant functions $\varphi : X \longrightarrow Y$ is a block. Indeed, if $(f, g) \in F \wr G$ and $(f, g)\Phi \cap \Phi \neq \emptyset$ then there exists $\varphi \in \Phi$ such that $(f, g)\varphi \in \Phi$. This implies that $[(f, g)\varphi](x) = f(x)\varphi(g^{-1}x)$ is constant as a function of x. Regularity forces f to be constant also, yielding $(f, g)\Phi = \Phi$.

We now prove that conditions (i) and (ii) imply that the action $F \wr G$ on Y^X is primitive. Let $\Phi \subseteq Y^X$ be a block for the action of $F \wr G$ on Y^X and suppose that $|\Phi| \geq 2$. Since the action of $F \wr G$ on Y^X is transitive, by virtue of Lemma 2.2.2 we may assume that Φ contains the constant function φ_0, where $\varphi_0(x) = y_0$ for all $x \in X$. By our assumptions, there exists $\varphi \in \Phi$ such that $\varphi \neq \varphi_0$. Thus we can find $x_1 \in X$ and $y_1 \in Y$, $y_1 \neq y_0$, satisfying $\varphi(x_1) = y_1$. Taking $f \in F^X$ such that

$$f(x)\varphi_0(x) = \varphi(x) \qquad \text{for all } x \in X,$$

we have that $(f, 1_G)\varphi_0 = \varphi$ and therefore

$$(f, 1_G)\Phi = \Phi \tag{2.21}$$

(because Φ is a block). In particular, for $u = f(x_1)$ we have $uy_0 = y_1$ and $u \notin H$. Since $N_F(H) = H$ (see Exercise 2.2.4) we can find $h \in H$ such that

$$u^{-1}hu \notin H.$$

Define $f_1, f_2 \in F^X$ by setting

$$f_1(x) = \begin{cases} h & \text{if } x = x_1 \\ 1_F & \text{otherwise} \end{cases}$$

and $f_2(x) = f_1(x)^{-1} f(x)^{-1} f_1(x) f(x)$, for all $x \in X$. Then we have

$$
f_2(x) = \begin{cases} h^{-1} u^{-1} h u \notin H & \text{if } x = x_1 \\ 1_F & \text{otherwise.} \end{cases}
$$

Since $(f_1, 1_G)\varphi_0 = \varphi_0$ we also have $(f_1, 1_G)\Phi = \Phi$ and therefore, recalling (2.21),

$$
(f_2, 1_G)\Phi \equiv (f_1, 1_G)^{-1}(f, 1_G)^{-1}(f_1, 1_G)(f, 1_G)\Phi = \Phi. \tag{2.22}
$$

Let us set $v = f_2(x_1)$. By virtue of Lemma 2.2.5, the set $B = \{\varphi(x_1) : \varphi \in \Psi_{x_1} \cap \Phi\}$ is a block for the action of F on Y. But B contains both y_0 (because $\varphi_0 \in \Psi_{x_1} \cap \Phi$) and $v y_0$ (because $(f_2, 1_G)\varphi_0 \in \Psi_{x_1} \cap \Phi$ and $[(f_2, 1_G)\varphi_0](x_1) = v y_0$); since $v y_0 \neq y_0$ (because $v \notin H$) and F is primitive on Y, necessarily $B \equiv Y$. It follows that Φ contains Ψ_{x_1}. Since $(1_F, g)\varphi_0 = \varphi_0$ it follows that $(1_F, g)\Phi = \Phi$, for all $g \in G$. Therefore from the elementary identity $(1_F, g)\Psi_{x_1} = \Psi_{g x_1}$ and the transitivity of the action of G on X, we deduce that

$$
\Phi \supseteq \bigcup_{x \in X} \Psi_x. \tag{2.23}
$$

Now we end the proof by showing that $\Phi = Y^X$. Indeed, any $\vartheta \in Y^X$ may be represented in the form

$$
\vartheta = \left[\prod_{x \in X} (f_x, 1_G) \right] \varphi_0,
$$

where $f_x \in F^X$ is defined by setting $f_x(x') = 1_F$ for $x' \neq x$ and by choosing $f_x(x)$ in such a way that $f_x(x) y_0 = \vartheta(x)$. Noticing that $(f_x, 1_G)\varphi_0 \in \Psi_x$, from (2.23) we deduce that $(f_x, 1_G)\varphi_0 \in \Phi$; since $\varphi_0 \in \Phi$ this implies that $(f_x, 1_G)\Phi = \Phi$ for all $x \in X$. Therefore

$$
\vartheta = \left[\prod_{x \in X} (f_x, 1_G) \right] \varphi_0 \in \left[\prod_{x \in X} (f_x, 1_G) \right] \Phi = \Phi.
$$

We have shown that $\Phi = Y^X$, so the exponentiation action of $F \wr G$ on Y^X is necessarily primitive. $\square$

2.3 Conjugacy classes of wreath products

This section takes its inspiration from the monograph of James and Kerber [38]. We recall that for any finite group G there exists a bijective correspondence between the set of conjugacy classes in G and the dual $\widehat{G}$ of G (that is, a complete set of pairwise inequivalent representations of G) (see [15, Corollary 1.3.16]).

2.3.1 A general description of conjugacy classes

Recall that, given distinct elements $x_1, x_2, \ldots, x_r$ in a set X, the permutation $c = (x_1 x_2 \cdots x_r)$ in $\mathrm{Sym}(X)$ that maps x_1 to x_2, x_2 to x_3, $\ldots$, x_{r-1} to x_r and x_r to x_1 and maps every other element of X to itself is called a *cycle*. The integer $r = \ell(c)$ is called the *length* of the cycle. Also, recalling that the *support* of a permutation $\pi \in \mathrm{Sym}(X)$ is the set consisting of all $x \in X$ such that $\pi(x) \neq x$, we have that given $\pi \in \mathrm{Sym}(X)$ there exists an integer $h \geq 0$ and cycles $c_1, c_2, \ldots, c_h$ with mutually disjoint supports such that $\pi = c_1 c_2 \cdots c_h$. This is called the *cycle decomposition* of π (and its expression is unique up to a permutation of the factors).

We now introduce a useful notation for cycles (see [59] and [15, Remark 3.1.1]). If $\pi \in \mathrm{Sym}(X)$ and c is a cycle of π, we write c in the form

$$c = (x \to \pi(x) \to \pi^2(x) \to \cdots \to \pi^{\ell(c)-1}(x) \to x), \qquad (2.24)$$

where x belongs to the support of c. In this way, if σ is another element of $\mathrm{Sym}(X)$ and we denote by σc the cycle in $\sigma \pi \sigma^{-1}$ corresponding to the cycle c of π as in (2.24) then, from the elementary identity $(\sigma \pi \sigma^{-1})^k[\sigma(x)] = \sigma \pi^k \sigma^{-1}[\sigma(x)] = \sigma \pi^k(x)$ for all $k \geq 0$, we get

$$\sigma c = (\sigma(x) \to \sigma \pi(x) \to \sigma \pi^2(x) \to \cdots \to \sigma \pi^{\ell(c)-1}(x) \to \sigma(x)). \quad (2.25)$$

Let now G be a finite group and, for $g \in G$, denote by $\mathfrak{C}(g) = \{h^{-1}gh : h \in G\}$ the *conjugacy class* of g in G. Suppose that G acts on a finite set X and denote by π this action: $\pi(g) \in \mathrm{Sym}(X)$ is the permutation of X associated with $g \in G$. Denote by $\mathcal{C}(\pi(g))$ the cycles of the permutation $\pi(g)$. Then any $c \in \mathcal{C}(\pi(g))$ is of the form (2.24):

$$c = (x \to \pi(g)x \to \cdots \to \pi(g)^{\ell(c)-1}x \to x),$$

where $x \in X$. The cycle decomposition of $\pi(g)$ is given by

$$\pi(g) = \prod_{c \in \mathcal{C}(\pi(g))} c \equiv \prod_{x \in \mathcal{O}(\pi(g))} (x \to \pi(g)x \to \cdots \to \pi(g)^{\ell(c)-1}x \to x),$$

where $\mathcal{O}(\pi(g)) \subset X$ denotes a set of representatives for the orbits of $\pi(g)$ on X. If $g, h \in G$ then $\mathcal{C}(\pi(hgh^{-1})) = h\mathcal{C}(\pi(g))$, where if $c = (x \to \pi(g)x \to \cdots \to \pi(g)^{\ell(c)-1}x \to x) \in \mathcal{C}(\pi(g))$ then $hc = (\pi(h)x \to \pi(h)\pi(g)x \to \cdots \to \pi(h)\pi(g)^{\ell(c)-1}x \to \pi(h)x)$ (cf. (2.25)).

Now let F be another finite group and denote by $\mathfrak{D} = \{\mathfrak{C}(f) : f \in F\}$ the set of its conjugacy classes. Form the wreath product $F \wr G = F^X \rtimes G$. In what follows, for the sake of simplicity we will use the notation gx to denote $\pi(g)x$. Also, for two elements a and b in a group H, we shall write $a \sim_H b$ if a and b are conjugate in H (that is, $\mathfrak{C}(a) = \mathfrak{C}(b)$). For $(f, g) \in F \wr G$ and $c = (x \to gx \to \cdots \to g^{\ell(c)-1}x \to x) \in C(\pi(g))$, we set

$$a_{c,x}(f, g) = f(g^{\ell(c)-1}x)f(g^{\ell(c)-2}x)\cdots f(gx)f(x) \in F. \qquad (2.26)$$

Lemma 2.3.1 *The conjugacy class of $a_{c,x}(f, g)$ in F does not depend on the particular expression (that is, the choice of the first element x) of the cycle $c = (x \to gx \to \cdots \to g^{\ell(c)-1}x \to x) \in C(\pi(g))$.*

Proof   If $(g^t x \to g^{t+1}x \to \cdots \to g^{t+\ell(c)-1}x \to g^t x) \in C(\pi(g))$ is another equivalent way to write c then

$$f(g^{t+\ell(c)-1}x)f(g^{t+\ell(c)-2}x)\cdots f(g^t x) \sim_F f(g^{\ell(c)-1}x)f(g^{\ell(c)-2}x)\cdots f(x),$$

because in any group the elements

$$a_1 a_2 \cdots a_{i-1} a_i a_{i+1} \cdots a_k \qquad \text{and} \qquad a_i a_{i+1} \cdots a_k a_1 a_2 \cdots a_{i-1}$$

are conjugate. $\square$

We now set $\mathcal{Y} = \{(\varphi, g) : g \in G \text{ and } \varphi : C(g) \to \mathfrak{D}\}$. In other words, $\mathcal{Y}$ consists of all pairs (φ, g) such that $g \in G$ and φ is a function which maps a cycle of the permutation on X associated with g to a conjugacy class of F. The group G acts on $\mathcal{Y}$ in a natural way: if $h \in G$ and $(\varphi, g) \in \mathcal{Y}$ then

$$h(\varphi, g) = (h\varphi, hgh^{-1}),$$

where $h\varphi : C(hgh^{-1}) \to \mathfrak{D}$ is defined by setting $h\varphi(c) = \varphi(h^{-1}c)$, for all $c \in C(g)$. By virtue of Lemma 2.3.1, the following definition is well posed.

Definition 2.3.2   Let $(f, g) \in F \wr G$ and define $\varphi : C(g) \to \mathfrak{D}$ by setting $\varphi(c)$ equal to the conjugacy class containing $a_{c,x}(f, g)$. We then denote by $A(f, g)$ the orbit of G on $\mathcal{Y}$ containing the element (φ, g).

Lemma 2.3.3   *Let $f \in F^X$, $g, h \in G$ and*

$$c = (x \to gx \to \cdots \to g^{\ell(c)-1}x \to x) \in C(g).$$

Then

$$a_{c,x}(f, g) = a_{hc,hx}(hf, hgh^{-1})$$

and

$$A(f, g) = A((1_F, h)(f, g)(1_F, h)^{-1}).$$

Proof   First recall that $hc = (hx \rightarrow hgx \rightarrow \cdots \rightarrow hg^{\ell(c)-1}x \rightarrow hx)$ is a cycle of hgh^{-1}. We then have

$$a_{hc,hx}(hf, hgh^{-1}) = f(h^{-1}hg^{\ell(c)-1}x) \cdots f(h^{-1}hx)$$
$$= f(g^{\ell(c)-1}x) \cdots f(x) = a_{c,x}(f, g).$$

Therefore, if for every $c \in \mathcal{C}(g)$ we denote by $\varphi(c)$ (resp. $\psi(hc)$) the conjugacy class of $a_{c,x}(f, g)$ (resp. $a_{hc,hx}(hf, hgh^{-1})$) then we have $h(\varphi, g) = (\psi, hgh^{-1})$ and $A(f, g) = A(hf, hgh^{-1}) = A((1_F, h)(f, g)(1_F, h)^{-1})$.   □

Lemma 2.3.4   *Let $f, f' \in F^X, g \in G$ and*

$$c = (x \rightarrow gx \rightarrow \cdots \rightarrow g^{\ell(c)-1}x \rightarrow x) \in \mathcal{C}(g).$$

Then

$$a_{c,x}(f, g) \sim_F a_{c,x}(f' f(gf')^{-1}, g),$$

where $(gf')^{-1}(y) = f'(g^{-1}y)^{-1}$ for all $y \in X$, and

$$A(f, g) = A((f', 1_G)(f, g)(f', 1_G)^{-1}).$$

Proof   We have

$$a_{c,x}(f' f(gf')^{-1}, g) = [f' f(gf')^{-1}](g^{\ell(c)-1}x)[f' f(gf')^{-1}](g^{\ell(c)-2}x)$$
$$\times \cdots \times [f' f(gf')^{-1}](x)$$
$$= f'(g^{\ell(c)-1}x) f(g^{\ell(c)-1}x)[f'(g^{\ell(c)-2}x)]^{-1}$$
$$\times f'(g^{\ell(c)-2}x) f(g^{\ell(c)-2}x)[f'(g^{\ell(c)-3}x)]^{-1}$$
$$\times \cdots \times f'(gx) f(gx)[f'(x)]^{-1} f'(x) f(x)[f'(g^{-1}x)]^{-1}$$
$$= f'(g^{\ell(c)-1}x) f(g^{\ell(c)-1}x) f(g^{\ell(c)-2}x)$$
$$\times \cdots \times f(gx) f(x)[f'(g^{-1}x)]^{-1}$$
$$= f'(g^{\ell(c)-1}x)a_{c,x}(f, g)[f'(g^{-1}x)]^{-1}$$
$$= f'(g^{-1}x)a_{c,x}(f, g)[f'(g^{-1}x)]^{-1}$$
$$\sim_F a_{c,x}(f, g).$$

Therefore, if for every $c \in \mathcal{C}(g)$ we denote by $\varphi(c)$ (resp. $\psi(c)$) the conjugacy class of $a_{c,x}(f, g)$ (resp. $a_{c,x}(f' f(gf')^{-1}, g)$), then $\varphi = \psi$ and $A(f, g) = A(f' f(gf')^{-1}, g) = A((f', 1_G)(f, g)(f', 1_G)^{-1})$.   □

Theorem 2.3.5   *Let $(f, g), (f', g') \in F \wr G$. Then we have $(f, g) \sim_{F \wr G} (f', g')$ if and only if $A(f, g) = A(f', g')$. In particular, the conjugacy classes of $F \wr G$ may be parameterized by the orbits of G on $\mathcal{Y}$.*

Proof   The "only if" part follows from Lemmas 2.3.3 and 2.3.4 after observing that if $(f'', g'') \in F \wr G$ then

$$(f'', g'')(f, g)(f'', g'')^{-1} = (f'', 1_G)(1_F, g'')(f, g)(1_F, g'')^{-1}(f'', 1_G)^{-1}.$$

Conversely, suppose that $A(f, g) = A(f', g')$. This implies the existence of $g'' \in G$ such that

$$g = g'' g' (g'')^{-1} \tag{2.27}$$

and

$$a_{c,x}(f, g) \sim_F a_{(g'')^{-1}c,(g'')^{-1}x}(f', g')$$

for all $c = (x \to gx \to \cdots \to g^{\ell(c)-1}x \to x) \in \mathcal{C}(g)$. From Lemma 2.3.3 we deduce that

$$A(f, g) = A(f', g') = A((1_F, g'')(f', g')(1_F, g'')^{-1}) = A(g'' f', g)$$

and

$$a_{c,x}(f, g) \sim_F a_{(g'')^{-1}c,(g'')^{-1}x}(f', g') = a_{c,x}(g'' f', g)$$

for all $c = (x \to gx \to \cdots \to g^{\ell(c)-1}x \to x) \in \mathcal{C}(g)$. Therefore, for every such $c \in \mathcal{C}(g)$ there exists $q_c \in F$ such that

$$f(g^{\ell(c)-1}x) f(g^{\ell(c)-2}x) \cdots f(x)$$
$$= q_c f'((g'')^{-1} g^{\ell(c)-1}x) f'((g'')^{-1} g^{\ell(c)-2}x) \cdots f'((g'')^{-1}x) q_c^{-1}. \tag{2.28}$$

Let $f'' : X \to F$ be defined as follows. For every $c = (x \to gx \to \cdots \to g^{\ell(c)-1}x \to x) \in \mathcal{C}(g)$ we recursively set $f''(g^{\ell(c)-1}x) = q_c$ and

$$f''(g^t x) = f(g^{t+1}x)^{-1} f''(g^{t+1}x) f'((g'')^{-1} g^{t+1}x) \tag{2.29}$$

for $t = \ell(c) - 2, \ell(c) - 3, \ldots, 1, 0$. It follows that

$$f(g^t x) = f''(g^t x) f'((g'')^{-1} g^t x) f''(g^{t-1}x)^{-1}$$

for $t = \ell(c) - 1, \ell(c) - 2, \ldots, 1$ and that, by (2.28),

$$f(x) = f''(x) f'((g'')^{-1}x) f''(g^{\ell(c)-1}x)^{-1}.$$

We deduce that

$$f(g^{\ell(c)-1}x)f(g^{\ell(c)-2}x)\cdots f(gx)f(x)$$
$$= [f''(g^{\ell(c)-1}x)f'((g'')^{-1}g^{\ell(c)-1}x)f''(g^{\ell(c)-2}x)^{-1}]$$
$$\times [f''(g^{\ell(c)-2}x)f'((g'')^{-1}g^{\ell(c)-2}x)f''(g^{\ell(c)-3}x)^{-1}]$$
$$\times \cdots \times [f''(gx)f'((g'')^{-1}gx)f''(x)^{-1}]$$
$$\times [f''(x)f'((g'')^{-1}x)f''(g^{\ell(c)-1}x)^{-1}]$$
$$=_* [f''(g^{\ell(c)-1}x)f'((g'')^{-1}g^{\ell(c)-1}x)f'((g'')^{-1}g^{\ell(c)-1}x)^{-1}$$
$$\times f''(g^{\ell(c)-1}x)^{-1}f(g^{\ell(c)-1}x)]$$
$$\times [f''(g^{\ell(c)-2}x)f'((g'')^{-1}g^{\ell(c)-2}x)f'((g'')^{-1}g^{\ell(c)-2}x)^{-1}$$
$$\times f''(g^{\ell(c)-2}x)^{-1}f(g^{\ell(c)-2}x)]$$
$$\times \cdots \times [f''(gx)f'((g'')^{-1}gx)f'((g'')^{-1}gx)^{-1}f''(gx)^{-1}f(gx)]$$
$$\times [f''(x)f'((g'')^{-1}x)f''(g^{\ell(c)-1}x)^{-1}]$$
$$= f(g^{\ell(c)-1}x)f(g^{\ell(c)-2}x)\cdots f(gx)[f''(x)f'((g'')^{-1}x)f''(g^{\ell(c)-1}x)^{-1}],$$
$$\tag{2.30}$$

where equality $=_*$ follows from (2.28). By comparing the first and last terms
of (2.30) we deduce that

$$f''(g^{-1}x) = f''(g^{\ell(c)-1}x) = f(x)^{-1}f''(x)f'((g'')^{-1}x). \tag{2.31}$$

It follows that $f = f''(g''f')(g(f'')^{-1})$ (compare this with (2.29) and (2.31))
and therefore, also using (2.27), we get

$$(f'', g'')(f', g')(f'', g'')^{-1} = (f''(g''f')(g(f'')^{-1}), g''g'(g'')^{-1}) = (f, g).$$

$$\square$$

2.3.2 Conjugacy classes of groups of the form $C_2 \wr G$

Let G be a finite group acting on a set X. Let $F = C_2 \equiv \{0, 1\}$ denote the
cyclic group of order two (we use additive notation). Then the wreath product
of C_2 by G (with respect to the action of G on X) is the set $C_2 \wr G = C_2^X \times
G = \{(\omega, g) : \omega \in C_2^X, g \in G)\}$ with the composition law $(\theta, g) \cdot (\omega, h) =
(\theta + g\omega, gh)$ for all $\theta, \omega \in C_2^X, g, h \in G$, where $g\omega(x) = \omega(g^{-1}x)$ for all
$x \in X$. The identity element is $(0_{C_2}, 1_G)$ and the inverse of $(\theta, g) \in C_2 \wr G$ is
given by $(\theta, g)^{-1} = (g^{-1}\theta, g^{-1})$.

Now, for $g \in G$, $c = (x \to gx \to \cdots \to g^{\ell(c)-1}x \to x) \in \mathcal{C}(g)$ and $\omega \in
C_2^X$, we have $a_{c,x}(\omega, g) = \omega(x) + \omega(gx) + \cdots + \omega(g^{\ell(c)-1}x)$. In particular, $a_{c,x}$

does not depend on x and so it will be denoted simply by $a_c(\omega, g)$. Moreover, $\mathcal{D}$ coincides with C_2, so that $\mathcal{Y} = \{(\varphi, g) : g \in G \text{ and } \varphi : \mathcal{C}(g) \to C_2\}$. We also recall that $A(\omega, g)$ denotes the orbit of G on $\mathcal{Y}$ containing (φ, g), where $\varphi(c) = a_c(\omega, g)$ for all $c \in \mathcal{C}(g)$.

Example 2.3.6 (The finite lamplighter group)   Let $G = C_n$ denote the cyclic group of order n: we shall use additive notation and identify G with $\mathbb{Z}/n\mathbb{Z}$. Thus we shall think of any element $k \in C_n$ as an integer representing the equivalence class $k + n\mathbb{Z}$. Let $X = C_n$ be equipped with the Cayley action of G. Note that, in our notation, $s(k\omega) = (s + k)\omega$.

Consider the wreath product $C_2 \wr C_n$. We denote by C_2^n the set of all maps $\theta : C_n \to C_2$. If $k \in C_n$ and $\theta \in C_2^n$ then $k\theta(j) = \theta(j - k)$ and the multiplication in $C_2 \wr C_n = \{(\theta, k) : \theta \in C_2^n, k \in C_n\}$ is given by

$$(\theta, k)(\omega, h) = (\theta + k\omega, k + h)$$

for all $(\theta, k), (\omega, h) \in C_2 \wr C_n$.

Now let $k \in C_n$ and denote by m the order of k. Then the (cyclic) group $\langle k \rangle$ generated by k is isomorphic to C_m and the cycles of k are the cosets of $\langle k \rangle$ in C_n. Setting $t = \frac{n}{m}$ we then have

$$\mathcal{C}(k) = \{(r \to r + k \to \cdots \to r + k(m - 1) \to r) : r = 0, 1, \ldots, t - 1\}.$$

We may identify $(r \to r + k \to \cdots \to r + k(m - 1) \to r)$ with r seen as an element of $C_t \cong \frac{C_n}{C_m}$ (that is, r is computed modulo t). Then the action of C_n on $\mathcal{C}(k)$ is the same thing as its action on $C_t \cong \frac{C_n}{C_m}$: for $j \in C_n$, the j-image of the cycle $(r \to r + k \to \cdots \to r + k(m - 1) \to r)$ is the cycle $(r + j \to r + j + k \to \cdots \to r + j + k(m - 1) \to r + j)$, which is determined by $r + j$ (computed modulo t). Note also that the conjugacy action of C_n on itself is trivial. Taking into account all these considerations, instead of $\mathcal{Y} = \{(\varphi, k) : k \in C_n \text{ and } \varphi : \mathcal{C}(k) \to C_2\}$ we will consider only the set $\coprod_{t | n} C_2^t$ (here C_2^t denotes, as usual, the set of all functions from C_t to C_2, and t varies among all divisors of n). For $(\omega, k) \in C_2 \wr C_n$, and for m and t as above, we denote by $\tilde{A}(\omega, k)$ the orbit of C_n on C_2^t containing the function $\varphi : C_t \to C_2$ defined by setting

$$\varphi(r) = \omega(r) + \omega(r + k) + \cdots + \omega(r + k(m - 1)) \qquad \text{for } r = 0, 1, \ldots, t - 1.$$

Recalling Definition 2.3.2, Theorem 2.3.5 immediately gives:

Theorem 2.3.7 *Two elements* $(\omega, k), (\theta, h) \in C_2 \wr C_n$ *are conjugate if and only if* $h = k$ *and* $\tilde{A}(\omega, k) = \tilde{A}(\theta, h)$.

Example 2.3.8 (The hyperoctahedral group)   Let $G = S_n$ denote the symmetric group of degree n. We recall some elementary facts on the conjugacy classes of S_n (see [11, 15]). If $\pi \in S_n$ and

$$\pi = (a_1 \to a_2 \to \cdots \to a_{\lambda_1} \to a_1)(b_1 \to b_2 \to \cdots \to b_{\lambda_2} \to b_1)$$
$$\cdots (c_1 \to c_2 \to \cdots \to c_{\lambda_k} \to c_1)$$

is the decomposition of π into disjoint cycles, with $\lambda_1 \geq \lambda_2 \geq \cdots \geq \lambda_k$ and $\lambda_1 + \lambda_2 + \cdots \lambda_k = n$ (trivial cycles are taken into account also) then the cycle structure of π is determined by the *partition* $\lambda = (\lambda_1, \lambda_2, \ldots, \lambda_k)$ of n. We write $\lambda \vdash n$ if λ is a partition of n. Let $\mathfrak{C}_\lambda \subset S_n$ be the set of all permutations whose cycle structure is equal to $\lambda \vdash n$. Then the sets $\mathfrak{C}_\lambda$, $\lambda \vdash n$, are precisely the conjugacy classes of S_n: two elements $\sigma, \pi \in S_n$ are conjugate if and only if they have the same cycle structure.

We now take $X = \{1, 2, \ldots, n\}$ and form the wreath product $C_2 \wr S_n = \{(\theta, \pi) : \pi \in S_n, \theta \in C_2^X\}$. Let $(\theta, \pi) \in C_2 \wr S_n$ and recall that $a_c(\theta, \pi) \equiv \theta(a_1) + \theta(a_2) + \cdots + \theta(a_{\ell(c)})$ for any cycle $c = (a_1 \to a_2 \to \cdots \to a_{\ell(c)} \to a_1)$ in $\mathcal{C}(\pi)$. For $i = 0, 1$ we then set $\mathcal{C}_{i,\theta}(\pi) = \{c \in \mathcal{C}(\pi) : a_c(\theta, \pi) = i\}$.

Definition 2.3.9

(i) A *double partition* of the positive integer n is a pair (λ, μ), where $\lambda \vdash k$, $\mu \vdash n - k$ and $0 \leq k \leq n$. We will write $(\lambda, \mu) \Vdash n$ to denote that (λ, μ) is a double partition of n.

(ii) If $(\lambda, \mu) \Vdash n$ we will denote by $\mathfrak{C}_{\lambda,\mu}$ the set of all pairs $(\theta, \pi) \in C_2 \wr S_n$ such that the cycle structures of the permutations

$$\prod_{\substack{c \in \mathcal{C}(\pi): \\ a_c(\theta,\pi)=0}} c \qquad \text{and} \qquad \prod_{\substack{c \in \mathcal{C}(\pi): \\ a_c(\theta,\pi)=1}} c$$

are equal to λ and μ, respectively.

In other words, λ (resp. μ) is determined by the lengths of the cycles $c = (a_1 \to a_2 \to \cdots \to a_{\ell(c)} \to a_1)$ of π on which $a_c(\theta, \pi)$ is equal to 0 (resp. 1).

Theorem 2.3.10   *The sets* $\mathfrak{C}_{\lambda,\mu}$, $(\lambda, \mu) \Vdash n$, *are precisely the conjugacy classes of* $C_2 \wr S_n$.

Proof   In the present setting $\mathcal{Y}$ is the set of all (φ, π) where $\pi \in S_n$ and $\varphi : C(\pi) \to C_2$. Given $(\varphi, \rho) \in \mathcal{Y}$ and $i = 0, 1$ we set

$$\rho_i = \prod_{\substack{c \in C(\rho): \\ \varphi(c)=i}} c.$$

Then (φ, π) and (ψ, σ) in $\mathcal{Y}$ are in the same S_n-orbit if and only if there exists $\eta \in S_n$ such that $\eta \pi \eta^{-1} = \sigma$ and $\varphi(\eta^{-1}c) = \psi(c)$ for all $c \in C(\pi)$, and this is in turn equivalent to the fact that π_i and σ_i have the same cyclic structure. Applying this fact to (θ, π) and $(\omega, \sigma) \in C_2 \wr S_n$ with $\varphi(c) = a_c(\theta, \pi)$ and $\psi(c) = a_c(\omega, \sigma)$, for all $c \in C(\pi)$, we conclude that (θ, π) and (ω, σ) are conjugate if and only if they belong to the same $\mathfrak{C}_{\lambda, \mu}$.   □

2.3.3   Conjugacy classes of groups of the form $F \wr S_n$

We now consider wreath products of the form $F \wr S_n$ with respect to the natural action of S_n on $X = \{1, 2, \ldots, n\}$. Therefore, $F \wr S_n = \{(f, \pi) : \pi \in S_n$ and $f : X \to F\}$ with the usual product law. We introduce a specific notation to parameterize the conjugacy classes of $F \wr S_n$. For $(f, \pi) \in F \wr S_n$, we consider the matrix

$$\alpha(f, \pi) = (\alpha_{\tau,k}(f, \pi))_{\substack{\tau \in \mathfrak{D} \\ k \in X}},$$

where $\alpha_{\tau,k}(f, \pi)$ equals the number of cycles $c \in C(\pi)$ such that $\ell(c) = k$ and $a_{c,x}(f, \pi) \in \tau$. Clearly,

$$\sum_{\tau \in \mathfrak{D}} \alpha_{\tau,k}(f, \pi) = \text{number of cycles of length } k \text{ in } \pi \qquad (2.32)$$

and

$$\sum_{k=1}^{n} k \sum_{\tau \in \mathfrak{D}} \alpha_{\tau,k}(f, \pi) = n. \qquad (2.33)$$

The matrix $\alpha(f, \pi)$ will be called the *type* of (f, π).

We denote by $\mathfrak{B} = \mathfrak{B}(F, n)$ the set of matrices

$$\beta - (\beta_{\tau,k})_{\substack{\tau \in \mathfrak{D} \\ k \in X}}$$

of nonnegative integers satisfying the condition

$$\sum_{k=1}^{n} k \sum_{\tau \in \mathfrak{D}} \beta_{\tau,k} = n. \qquad (2.34)$$

Theorem 2.3.11 *Two elements in $F \wr S_n$ are conjugate if and only if they have the same type. Moreover, the set of conjugacy classes of $F \wr S_n$ is in bijective correspondence with $\mathfrak{B}$.*

Proof Let us show that two elements (f, π) and (f', π') in $F \wr S_n$ are conjugate if and only if the have the same type. Observe that, by virtue of Theorem 2.3.5, it suffices to prove that $A(f, \pi) = A(f', \pi')$ if and only if $\alpha(f, \pi) = \alpha(f', \pi')$. The "only if" part is obvious.

Conversely suppose that $\alpha(f, \pi) = \alpha(f', \pi')$. From (2.32) we get $\pi \sim_{S_n} \pi'$, because π and π' have the same cyclic structure. Moreover, we can choose $\sigma \in \mathfrak{S}_n$ such that $\sigma \pi \sigma^{-1} = \pi'$ and, in addition, $a_{c,x}(f, \pi) \sim_F a_{\sigma c, \sigma x}(f', \pi')$ for all $c \in C(\pi)$ and for $x \in \mathcal{O}(c)$, the orbit of c in $X = \{1, 2, \ldots, n\}$. Indeed, to construct a permutation $\sigma \in S_n$ such that $\sigma \pi \sigma^{-1}$, the first step is to choose a bijection between $C(\pi)$ and $C(\pi')$ that preserves the length of the cycles. Since $\alpha(f, \pi) = \alpha(f', \pi')$ we can choose that bijection in such a way that, for every $c \in C(\pi)$, the elements $a_{c,x}(f, \pi)$ and $a_{\sigma c, \sigma x}(f', \pi')$ belong to the same conjugacy class. This ensures that $A(f, \pi) = A(f', \pi')$ and proves the first statement of the theorem.

To complete the proof, we note that $\alpha(f, \pi) \in \mathfrak{B}$ for all $(f, \pi) \in F \wr S_n$. Let now show that given any $\beta \in \mathfrak{B}$ there exists $(f, \pi) \in F \wr S_n$ such that $\alpha(f, \pi) = \beta$. We first take $\pi \in S_n$, which has $\sum_{\tau \in \mathfrak{D}} \beta_{\tau, k}$ many cycles of length k and then construct f as follows. We arbitrarily make the partition

$$C(\pi) = \coprod_{\substack{\tau \in \mathfrak{D} \\ k \in X \\ \beta_{\tau, k} \neq 0}} C_{\tau, k}(\pi)$$

in such a way that $|C_{\tau, k}(\pi)| = \beta_{\tau, k}$ for all $\tau \in \mathfrak{D}$ and $k \in X$. Then, for each $\tau \in \mathfrak{D}$, $k \in X$ and $c \in C_{\tau, k}(\pi)$, we arbitrarily select an element $x = x(c) \in \mathcal{O}(c)$ and one element $f_0 \in \tau$ and then set

$$f(y) = \begin{cases} f_0 & \text{if } y = x \\ 1_F & \text{otherwise} \end{cases}$$

for all $y \in \mathcal{O}(c)$. Since the orbits $\mathcal{O}_c$, $c \in C(\pi)$, form a partition of X this defines an element $f \in F^X$. It immediately follows by construction that $\alpha(f, \pi) = \beta$.

This shows that the map $(f, \pi) \mapsto \alpha(f, \pi)$ induces a bijection between the set of conjugacy classes of $F \wr S_n$ and $\mathfrak{B}$. $\square$

As a consequence of the above theorem, we have that the number of conjugacy classes of $F \wr S_n$ is equal to $|\mathfrak{B}|$. We now give a formula explicitly expressing such a number. To this purpose we denote by $c(n, h)$ the set of all finite sequences $(a_1, a_2, \ldots, a_h)$, where $a_1, a_2, \ldots, a_h$ are nonnegative integers and $a_1 + a_2 + \cdots + a_h = n$ (such a sequence is called a *composition* of n of length h). Moreover, we denote by $p(n)$ the number of partitions of n.

Corollary 2.3.12   *The number of conjugacy classes of $F \wr S_n$ is equal to*

$$\sum_{(a_1, a_2, \ldots, a_h) \in c(n,h)} p(a_1) p(a_2) \cdots p(a_h), \qquad (2.35)$$

where $h = |\mathfrak{D}|$.

Proof   Let us set

$$\Sigma(F, n) = \left\{ a = (a_\tau)_{\tau \in \mathfrak{D}} : a_\tau \geq 0, \sum_{\tau \in \mathfrak{D}} a_\tau = n \right\} \qquad (2.36)$$

and, for $a \in \Sigma(F, n)$,

$$\Lambda(a) = \{\lambda^{(a)} = (\lambda_\tau)_{\tau \in \mathfrak{D}} : \lambda_\tau \vdash a_\tau \text{ for all } \tau \in \mathfrak{D}\}. \qquad (2.37)$$

It is obvious that the map

$$\beta \mapsto \left(a(\beta), \lambda^{(a(\beta))} \right) \qquad (2.38)$$

defined by

$$a(\beta)_\tau = \sum_{k=1}^{n} k \beta_{\tau, k}$$

and

$$\lambda_\tau^{(a(\beta))} = (n^{\beta_{\tau,n}}, (n-1)^{\beta_{\tau,n-1}}, \cdots, 1^{\beta_{\tau,1}})$$

for all $\beta = (\beta_{\tau,k})_{\tau \in \mathfrak{D}, k \in X} \in \mathfrak{B}$ and $\tau \in \mathfrak{D}$ establishes a bijection between $\mathfrak{B}$ and the set

$$\mathfrak{N} = \mathfrak{N}(F, n) = \{(a, \lambda^{(a)}) : a \in \Sigma(F, n), \lambda^{(a)} \in \Lambda(a)\}. \qquad (2.39)$$

It is now obvious that $|\mathfrak{N}|$ equals the quantity expressed in (2.35).   □

Example 2.3.13 (Conjugacy classes of $S_m \wr S_n$)   We now consider the conjugacy classes of $S_m \wr S_n$, that is, we specify the above analysis for $F \wr S_n$ with $F = S_m$. Recall that for each $\mu \vdash m$ we denote by $\mathfrak{C}_\mu$ the set of all permutations in S_m whose cycle structure is equal to μ. We set

$$\Sigma(m, n) = \left\{ \nu = (n_\mu)_{\mu \vdash m} : n_\mu \geq 0, \sum_{\mu \vdash m} n_\mu = n \right\} \qquad (2.40)$$

and, for $\nu \in \Sigma(m, n)$,

$$\Lambda(\nu) = \{\lambda^{(\nu)} = (\lambda_\mu)_{\mu \vdash m} : \lambda_\mu \vdash n_\mu\}. \tag{2.41}$$

Note that $\Sigma(m, n)$ is nothing other than $\Sigma(S_m, n)$, as in (2.36), and $\Lambda(\nu)$ corresponds to $\Lambda(a)$ as in (2.37). It follows from Corollary 2.3.12 and its proof that the conjugacy classes of $S_m \wr S_n$ are parameterized by the set

$$\mathfrak{N} = \mathfrak{N}(m, n) = \left\{ \left(\nu, \lambda^{(\nu)} \right) : \nu \in \Sigma(m, n), \lambda^{(\nu)} \in \Lambda(\nu) \right\}. \tag{2.42}$$

2.4 Representation theory of wreath products

This section takes its inspiration from the monographs by James and Kerber [38] and Huppert [35].

2.4.1 The irreducible representations of wreath products

Let G, X and F be as in the previous sections. Now we want to develop the representation theory of the wreath product $F \wr G$. It is well known (see, for instance, Theorem 9.1.6 and Corollary 9.1.7 in [11]) that every irreducible representation σ of the base group F^X is of the form

$$\sigma = \bigotimes_{x \in X} \sigma_x,$$

where $\sigma_x \in \widehat{F}$ for all $x \in X$. Recalling that $F^X \cong F^X \times \{1_G\} \subset F \wr G$, for $f_0 \in F^X$ we set

$$\sigma(f_0, 1_G) = \bigotimes_{x \in X} \sigma_x(f_0(x)).$$

In other words, $V_\sigma = \bigotimes_{x \in X} V_{\sigma_x}$, where V_{σ_x} denotes the representation space of σ_x, and we have

$$[\sigma(f_0, 1_G)] \left(\bigotimes_{x \in X} v_x \right) = \bigotimes_{x \in X} \sigma_x(f_0(x)) v_x.$$

Recalling that $F^X \cong F^X \times \{1_G\}$ is normal in $F \wr G$ we have the following:

Lemma 2.4.1 *Let $(f, g) \in F \wr G$ and $\sigma_x \in \widehat{F}$, for each $x \in X$. Then the (f, g)-conjugate of $\sigma = \bigotimes_{x \in X} \sigma_x$ is given by the formula*

$$^{(f,g)}\sigma = \bigotimes_{x \in X} {}^{f(x)}\sigma_{g^{-1}x} \sim \bigotimes_{x \in X} \sigma_{g^{-1}x}.$$

Proof   Fix $f_0 \in F^X$. Since

$$(f, g)^{-1}(f_0, 1_G)(f, g) = (g^{-1}f^{-1}, g^{-1})(f_0, 1_G)(f, g)$$
$$= (g^{-1}f^{-1} \cdot g^{-1}(f_0 f), 1_G),$$

we have on the one hand

$$^{(f,g)}\sigma(f_0, 1_G) = \sigma[(f, g)^{-1}(f_0, 1_G)(f, g)]$$
$$= \sigma(g^{-1}f^{-1} \cdot g^{-1}(f_0 f), 1_G)$$
$$= \bigotimes_{x \in X} \sigma_x(f(gx)^{-1} f_0(gx) f(gx))$$
$$= \bigotimes_{x \in X} \sigma_{g^{-1}x}(f(x)^{-1} f_0(x) f(x))$$
$$= \bigotimes_{x \in X} {}^{f(x)}\sigma_{g^{-1}x}(f_0(x))$$
$$= \left[\bigotimes_{x \in X} {}^{f(x)}\sigma_{g^{-1}x} \right] (f_0, 1_G),$$

that is,

$$^{(f,g)}\sigma = \bigotimes_{x \in X} {}^{f(x)}\sigma_{g^{-1}x}.$$

On the other hand, since $f(x) \in F$ and $\sigma_{g^{-1}x} \in \widehat{F}$ we have

$$^{f(x)}\sigma_{g^{-1}x} \sim \sigma_{g^{-1}x},$$

so that

$$^{(f,g)}\sigma \sim \bigotimes_{x \in X} \sigma_{g^{-1}x}.$$

$\square$

Lemma 2.4.2   *For every* $\sigma = \left(\bigotimes_{x \in X} \sigma_x \right) \in \widehat{F^X}$, *the inertia group of* σ *with respect to* $F \wr G$ *is given by*

$$I_{F \wr G}(\sigma) = F \wr T_G(\sigma),$$

where

$$T_G(\sigma) = \{g \in G : \sigma_{gx} \sim \sigma_x \text{ for all } x \in X\}$$
$$\equiv \{g \in G : \sigma_{gx} = \sigma_x \text{ for all } x \in X\} \qquad (2.43)$$

(recall that $\widehat{F^X}$ denotes a set of representatives of the irreducible representations of F^X).

Proof From Lemma 2.4.1 we immediately deduce that

$$I_{F \wr G}(\sigma) = \{(f, g) : \sigma_{gx} \sim \sigma_x \text{ for all } x \in X\} = F^X \rtimes T_G(\sigma) = F \wr T_G(\sigma).$$

□

Lemma 2.4.3 *Every* $\sigma = \left(\bigotimes_{x \in X} \sigma_x\right) \in \widehat{F^X}$ *has an extension* $\tilde{\sigma}$, *to the whole of* $I_{F \wr G}(\sigma)$, *defined by setting*

$$\tilde{\sigma}(f, g) \left(\bigotimes_{x \in X} v_x\right) = \bigotimes_{x \in X} \sigma_{g^{-1}x}(f(x))v_{g^{-1}x}$$

for all $(f, g) \in F \wr T_G(\sigma)$ *and* $\bigotimes_{x \in X} v_x \in \bigotimes_{x \in X} V_{\sigma_x}$.

Proof From the definitions of $\tilde{\sigma}$ and $T_G(\sigma)$ we immediately have

$$\tilde{\sigma}(f, g)\left(\bigotimes_{x \in X} v_x\right) = \bigotimes_{x \in X} \sigma_{g^{-1}x}(f(x))v_{g^{-1}x} = \bigotimes_{x \in X} \sigma_x(f(x))v_{g^{-1}x}.$$

We now prove that $\tilde{\sigma}$ is a homomorphism. Let $(f_1, g_1), (f_2, g_2) \in F \wr T_G(\sigma)$. We have

$$\tilde{\sigma}\left((f_1, g_1) \cdot (f_2, g_2)\right)\left(\bigotimes_{x \in X} v_x\right) = \tilde{\sigma}\left((f_1 \cdot (g_1 f_2), g_1 g_2)\right)\left(\bigotimes_{x \in X} v_x\right)$$
$$= \bigotimes_{x \in X} \sigma_x(f_1(x) f_2(g_1^{-1}x))v_{g_2^{-1}g_1^{-1}x}$$

and

$$\tilde{\sigma}((f_1, g_1)\left(\tilde{\sigma}(f_2, g_2)\left(\bigotimes_{x \in X} v_x\right)\right) = \tilde{\sigma}(f_1, g_1)\left(\bigotimes_{x \in X} \sigma_x(f_2(x))v_{g_2^{-1}x}\right)$$
$$= \bigotimes_{x \in X} \sigma_x(f_1(x))\sigma_{g_1^{-1}x}(f_2(g_1^{-1}x))v_{g_2^{-1}g_1^{-1}x}$$
$$= \bigotimes_{x \in X} \sigma_x(f_1(x))\sigma_x(f_2(g_1^{-1}x))v_{g_2^{-1}g_1^{-1}x}$$
$$= \bigotimes_{x \in X} \sigma_x(f_1(x) f_2(g_1^{-1}x))v_{g_2^{-1}g_1^{-1}x}.$$

That is, $\tilde{\sigma}\left((f_1, g_1) \cdot (f_2, g_2)\right) = \tilde{\sigma}(f_1, g_1)\tilde{\sigma}(f_2, g_2)$ and this proves that $\tilde{\sigma}$ is a representation.

□

Table 2.1

General situation	Present setting
G	$F \wr G$
N	F^X
$I_G(\sigma)$	$F \wr T_G(\sigma) = F^X \rtimes T_G(\sigma)$
$I_G(\sigma)/N$	$(F^X \rtimes T_G(\sigma))/F^X = T_G(\sigma)$

We now apply the little group method (see Section 1.3.2) in order to describe all the irreducible representations of the wreath product $F \wr G$. Table 2.1 gives the correspondence between the notation in the general situation and that in the present setting.

Let Σ be a system of representatives for the $(F \wr G)$-conjugacy classes of $\widehat{F^X}$. For each $\sigma \in \Sigma$, denote by $\widetilde{\sigma}$ its extension to $I_{F \wr G}(\sigma)$ (see Lemma 2.4.3). Moreover, for each $\psi \in \widehat{T_G}(\sigma)$ denote by $\overline{\psi}$ its inflation (1.71) to $I_{F \wr G}(\sigma)$ (using the homomorphism $I_{F \wr G}(\sigma) \to T_G(\sigma) \cong I_{F \wr G}(\sigma)/F^X$). This means that

$$\overline{\psi}(f, g) = \psi(g)$$

for all $(f, g) \in I_{F \wr G}(\sigma) \equiv F \wr T_G(\sigma)$. Then, an application of Theorem 1.3.11 yields the following:

Theorem 2.4.4 *With the above notation, we have*

$$\widehat{F \wr G} = \{\mathrm{Ind}_{I_{F \wr G}(\sigma)}^{F \wr G}(\widetilde{\sigma} \otimes \overline{\psi}) : \sigma \in \Sigma, \psi \in \widehat{T_G}(\sigma)\},$$

that is, the right-hand side constitutes a complete list of pairwise inequivalent irreducible representations of $F \wr G$.

2.4.2 The character and matrix coefficients of the representation $\widetilde{\sigma}$

When expressing the character and matrix coefficients of one of the irreducible representations referred to in Theorem 2.4.4, the main problem is to compute the character and the matrix coefficients of $\widetilde{\sigma}$. Indeed, the matrix coefficients of $\overline{\psi}$ can be obtained by composing those of ψ with the homomorphism $I_{F \wr G}(\sigma) \to T_G(\sigma) \cong I_{F \wr G}(\sigma)/F^X$. For $\widetilde{\sigma} \otimes \overline{\psi}$ we can use the well-known formulas for the character and matrix coefficients of a tensor product (if (ρ_1, V_1) and (ρ_2, V_2) are two irreducible representations of a group G then

$\chi^{\rho_1 \otimes \rho_2}(g) = \chi^{\rho_1}(g)\chi^{\rho_2}(g)$ and $u^{\rho_1 \otimes \rho_2}_{i,j;k,\ell}(g) = u^{\rho_1}_{i,k}(g)u^{\rho_2}_{j,\ell}(g)$ for all $g \in G$.
Finally, we may apply to $\mathrm{Ind}^G_{I_F G(\sigma)}(\widetilde{\sigma} \otimes \widetilde{\psi})$ the formulas (1.14) and (1.7) for an induced representation. Therefore, in this section we focus our attention on $\widetilde{\sigma}$.

Recall that G acts on the finite set X. Let $\sigma = \otimes_{x \in X}\sigma_x$, where $\sigma_x \in \widehat{F}$ for all $x \in X$. Recalling the definition of the subgroup $T_G(\sigma) \leq G$, we observe that if $\Omega_1, \Omega_2, \ldots, \Omega_m \subset X$ denote the corresponding $T_G(\sigma)$-orbits then for $x, y \in X$ we have $\sigma_x = \sigma_y$ if and only if there exists $1 \leq i \leq m$ such that $x, y \in \Omega_i$. Therefore there exist pairwise inequivalent irreducible representations $\sigma_1, \sigma_2, \ldots, \sigma_m \in \widehat{F}$ such that $\sigma_x = \sigma_i$ for all $x \in \Omega_i$, $i = 1, 2, \ldots, m$. In particular, we have

$$\sigma = \bigotimes_{i=1}^{m} \bigotimes_{x \in \Omega_i} \sigma_x.$$

Let $v^i_1, v^i_2 \ldots, v^i_{d_i}$ be an orthonormal basis in V_{σ_i}, where $d_i = \dim V_{\sigma_i}$, $i = 1, 2, \ldots, m$. Then the character and the matrix coefficients of σ_i are given by

$$\chi_i(t) = \sum_{j=1}^{d_i} \langle \sigma(t)v^i_j, v^i_j \rangle_{V_i}$$

and

$$u^i_{j,k}(t) = \langle \sigma(t)v^i_k, v^i_j \rangle_{V_i}$$

for all $t \in F$ and $j, k = 1, 2, \ldots, d_i$, respectively. Moreover, they satisfy the following elementary properties:

$$\sum_{k=1}^{d_i} u^i_{j,k}(t)u^i_{k,h}(s) = u^i_{j,h}(ts) \qquad \text{and} \qquad \sigma(t)v^i_k = \sum_{j=1}^{d_i} u^i_{j,k}(t)v^i_j.$$

For $g \in T_G(\sigma)$, denote by $\mathcal{C}_i(g)$ the set of cycles of the permutation induced by g on Ω_i. We also denote by $\mathcal{A}$ the set of all maps $\varphi : X \to \mathbb{N}$ such that $\varphi(x) \in \{1, 2, \ldots, d_i\}$ for all $x \in \Omega_i$ and $i = 1, 2, \ldots, m$. Then, for every $\varphi \in \mathcal{A}$ we set

$$v_\varphi = \bigotimes_{i=1}^{m} \bigotimes_{x \in \Omega_i} v^i_{\varphi(x)}.$$

It is clear that $\{v_\varphi : \varphi \in \mathcal{A}\}$ is an orthonormal basis for $V_\sigma = \otimes_{x \in X} V_{\sigma_x}$. In the following, we shall use the notation $a_{c,x}(f, g)$ from (2.26) for $(f, g) \in F \wr G$, $x \in X$ and $c \in \mathcal{C}_i(g)$.

Theorem 2.4.5   *The matrix coefficients and the character of the extension $\tilde{\sigma}$ of σ are given respectively by*

$$u^{\tilde{\sigma}}_{\psi,\varphi}(f,g) = \prod_{i=1}^{m} \prod_{x\in\Omega_i} u^i_{\psi(x),\varphi(g^{-1}x)}(f(x)), \qquad (2.44)$$

where $\varphi, \psi \in \mathcal{A}$, and

$$\chi_{\tilde{\sigma}}(f,g) = \prod_{i=1}^{m} \prod_{c\in\mathcal{C}_i(g)} \chi_{\sigma_i}(a_c(f,g)) \qquad (2.45)$$

for all $(f,g) \in I_{F\wr G}(\sigma)$.

Proof   From Lemma 2.4.3 we get, for $(f,g) \in I_{F\wr G}(\sigma)$,

$$\tilde{\sigma}(f,g)v_\varphi = \bigotimes_{i=1}^{m} \bigotimes_{x\in\Omega_i} \sigma_x(f(x))v^i_{\varphi(g^{-1}x)}$$

$$= \bigotimes_{i=1}^{m} \bigotimes_{x\in\Omega_i} \left(\sum_{j=1}^{d_i} u^i_{j,\varphi(g^{-1}x)}(f(x))v^i_j \right)$$

$$= \sum_{\psi\in\mathcal{A}} \left(\prod_{i=1}^{m} \prod_{x\in\Omega_i} u^i_{\psi(x),\varphi(g^{-1}x)}(f(x)) \right) v_\psi$$

and this proves (2.44). Similarly, starting from (2.45), we can find the character of $\tilde{\sigma}$:

$$\chi_{\tilde{\sigma}}(f,g) = \sum_{\varphi\in\mathcal{A}} \langle \tilde{\sigma}(f,g)v_\varphi, v_\varphi \rangle$$

$$= \sum_{\varphi\in\mathcal{A}} \left(\prod_{i=1}^{m} \prod_{x\in\Omega_i} u^i_{\varphi(x),\varphi(g^{-1}x)}(f(x)) \right)$$

$$= \sum_{\varphi\in\mathcal{A}} \left(\prod_{i=1}^{m} \prod_{x\in\Omega_i} u^i_{\varphi(gx),\varphi(x)}(f(gx)) \right)$$

$$= \sum_{\varphi\in\mathcal{A}} \left(\prod_{i=1}^{m} \prod_{c\in\mathcal{C}_i(g)} u^i_{\varphi(gx),\varphi(x)}(f(gx))u^i_{\varphi(g^2x),\varphi(gx)}(f(g^2x)) \right.$$

$$\left. \times \cdots \times u^i_{\varphi(g^{\ell(c)-1}x),\varphi(g^{\ell(c)-2}x)}(f(g^{\ell(c)-1}x))u^i_{\varphi(x),\varphi(g^{\ell(c)-1}x)}(f(x)) \right)$$

$$
=\prod_{i=1}^{m}\prod_{c\in\mathcal{C}_i(g)}\sum_{\varphi(gx)=1}^{d_i}\sum_{\varphi(g^2x)=1}^{d_i}\cdots\sum_{\varphi(g^{\ell(c)-1}x)=1}^{d_i}\sum_{\varphi(x)=1}^{d_i}u^i_{\varphi(gx),\varphi(x)}(f(gx))
$$

$$
\times u^i_{\varphi(g^2x),\varphi(gx)}(f(g^2x))\cdots u^i_{\varphi(g^{\ell(c)-1}x),\varphi(g^{\ell(c)-2}x)}(f(g^{\ell(c)-1}x))
$$

$$
\times u^i_{\varphi(x),\varphi(g^{\ell(c)-1}x)}(f(x))
$$

$$
=\prod_{i=1}^{m}\prod_{c\in\mathcal{C}_i(g)}\sum_{\varphi(x)=1}^{d_i}u^i_{\varphi(x),\varphi(x)}(a_{c,x}(f,g))
$$

$$
=\prod_{i=1}^{m}\prod_{c\in\mathcal{C}_i(g)}\chi_{\sigma_i}(a_c(f,g)),
$$

where $c=(x\to gx\to\cdots\to g^{\ell(c)-1}x\to x)\in\mathcal{C}_i(g)$.   □

As a particular case of Theorem (2.4.5) we have

Corollary 2.4.6 *Suppose that $\sigma_x=\sigma$ for all $x\in X$ (so that $T_G(\sigma)=G$ and $\widetilde{\sigma}\in\widehat{F\wr G}$). Then we have*

$$
\chi_{\widetilde{\sigma}}(f,g)=\prod_{c\in\mathcal{C}(g)}\chi_\sigma(a_c(f,g))
$$

for all $(f,g)\in F\wr G$. In particular, for all $g\in G$ and $f\in F^X$,

$$
\chi_{\widetilde{\sigma}}(1_F,1_G)=\dim(V_\sigma)^{|X|},
$$

$$
\chi_{\widetilde{\sigma}}(1_F,g)=\dim(V_\sigma)^{|\mathcal{C}(g)|},
$$

$$
\chi_{\widetilde{\sigma}}(f,1_G)=\prod_{x\in X}\chi_\sigma(f(x)).
$$

Finally, if f is constant, say $f(x)=t\in F$ for all $x\in X$, and for $1\le k\le|X|$ we set $a_k(g)=|\{c\in\mathcal{C}(g):\ell(c)=k\}|$, then we have

$$
\chi_{\widetilde{\sigma}}(f,g)=\prod_{k=1}^{|X|}\chi_\sigma(t^k)^{a_k(g)}.
$$

        □

2.5 Representation theory of groups of the form $C_2\wr G$

Let G be a finite group acting transitively on a finite set X. For $\omega,\theta\in C_2^X$ we set $\omega\cdot\theta=\sum_{x\in X}\omega(x)\theta(x)\in C_2$. Define the character χ_θ of C_2^X by setting $\chi_\theta(\omega)=(-1)^{\omega\cdot\theta}$. Then the dual group of C_2^X is just $\widehat{C_2^X}=\{\chi_\theta:\theta\in C_2^X\}$

and G acts on it in a natural way: for all $g \in G$ and $\omega, \theta \in C_2^X$ one defines $g\chi_\theta(\omega) = \chi_\theta(g^{-1}\omega)$, that is, $g\chi_\theta = \chi_{g\theta}$. The action of G on $\widehat{C_2^X}$ is equivalent to the action on C_2^X and both can be identified with the action on the subsets of X. In particular, for $\theta \in C_2^X$ the stabilizer $G_\theta = \{g \in G : g\chi_\theta = \chi_\theta\}$ (we use this notation in place of $T_G(\theta)$) coincides with the stabilizer of the subset $X_\theta = \{x \in X : \theta(x) = 0\}$. Then the extension $\tilde{\chi}_\theta \in \widehat{C_2 \wr G_\theta}$ of the character χ_θ is given by $\tilde{\chi}_\theta(\omega, g) = \chi_\theta(\omega)$ for all $\omega \in C_2^X$ and $g \in G_\theta$. Moreover, if $\eta \in \widehat{G_\theta}$ then its inflation $\bar{\eta}$ to $C_2 \wr G_\theta$ is given by $\bar{\eta}(\omega, g) = \eta(g)$ for all $\omega \in C_2^X$ and $g \in G_\theta$. Both $\tilde{\chi}_\theta$ and $\bar{\eta}$ are irreducible $(C_2 \wr G_\theta)$-representations and so is their tensor product $\tilde{\chi}_\theta \otimes \bar{\eta}$: clearly

$$\tilde{\chi}_\theta \otimes \bar{\eta}(\omega, g) = \chi_\theta(\omega)\eta(g), \tag{2.46}$$

and χ_θ is one dimensional. Applying Theorem 2.4.4 we deduce the following:

Theorem 2.5.1 *Let Θ be a system of representatives for the orbits of G on C_2^X (that is, any G-orbit has exactly one element in Θ). Then*

$$\widehat{C_2 \wr G} = \left\{ \mathrm{Ind}_{C_2 \wr G_\theta}^{C_2 \wr G}(\tilde{\chi}_\theta \otimes \bar{\eta}) : \theta \in \Theta \text{ and } \eta \in \widehat{G_\theta} \right\},$$

that is, the right-hand side constitutes a complete list of pairwise inequivalent irreducible representations of $C_2 \wr G$. □

2.5.1 Representation theory of the finite lamplighter group $C_2 \wr C_n$

The group $C_2 \wr C_n$ has already been introduced in Example 2.3.6. Every irreducible representation of C_n is one dimensional; it can be identified with its character and we have $\widehat{C_n} = \{e_k : k \in C_n\}$, where $e_k(h) = \exp\left(2\pi i \frac{hk}{n}\right)$ for all $h, k \in C_n$.

We may think of an element $\theta \in C_2^n$ as a periodic function $\theta : \mathbb{Z} \to C_2$ satisfying $\theta(k + n) = \theta(k)$ for any $k \in \mathbb{Z}$. Recall that the *period* of θ is the smallest positive integer $t = t(\theta)$ such that $\theta(k + t) = \theta(k)$ for any $k \in \mathbb{Z}$. It is easy to show that t divides n; morover, if $n = mt$ then the stabilizer of θ is the subgroup $C_m = \langle t \rangle$ (recall also that, for any divisor m of n, the subgroup of C_n isomorphic to C_m is unique [50]). The characters of the subgroup $\langle t \rangle$ are $e_0|_{\langle t \rangle}, e_1|_{\langle t \rangle}, \ldots, e_{m-1}|_{\langle t \rangle}$, where $e_0, e_1, \ldots, e_{m-1}$ are as above. Indeed, for $0 \leq r, \ell \leq m - 1$ we have $e_r(\ell t) = \exp\left(2\pi i \frac{r\ell t}{n}\right) = \exp\left(2\pi i \frac{r\ell}{m}\right)$. We set $e_r|_{\langle t \rangle}(k) = e_r(k)$ when $k \in \langle t \rangle$ and $e_r|_{\langle t \rangle}(k) = 0$ otherwise. We also set $m = m(\theta) = \frac{n}{t(\theta)}$, but we shall simply write t and m when the element θ is clear from the context.

For $\theta \in C_2^n$ and $0 \le r \le m-1$, the character $\widetilde{\chi}_\theta \otimes \overline{e_r|_{\langle t \rangle}}$ of $C_2^n \wr \langle t \rangle$ is given by

$$\widetilde{\chi}_\theta \otimes \overline{e_r|_{\langle t \rangle}}(\omega, \ell t) = \chi_\theta(\omega)e_r(\ell t)$$

for all $\omega \in C_2^n$ and $\ell = 0, 1, \ldots, m-1$. Let Θ be a set of representatives for the orbits of C_n on C_2^n (such orbits may be enumerated by mean of the so-called Polya–Redfield theory; see [54] for an elementary account and [44] for a more comprehensive treatment). Then we may apply Theorem 2.5.1.

Theorem 2.5.2   *We have*

$$\widehat{C_2^n \wr C_n} = \left\{ \mathrm{Ind}_{C_2 \wr \langle t(\theta) \rangle}^{C_2 \wr C_n} \left[\widetilde{\chi}_\theta \otimes \overline{e_r|_{\langle t(\theta) \rangle}} \right] : \theta \in \Theta, r = 0, 1, \ldots, m(\theta) - 1 \right\},$$

that is, the right-hand side constitutes a complete list of pairwise inequivalent irreducible representations of $C_2^n \wr C_n$. $\square$

2.5.2 Representation theory of the hyperoctahedral group $C_2 \wr S_n$

As in Example 2.3.8, we have $G = S_n$ and $X = \{1, 2, \ldots, n\}$. For any $0 \le k \le n$, choose $\theta^{(k)} \in C_2^X$ such that $|\{j \in X : \theta^{(k)}(j) = 0\}| = k$. Then the set $\{\theta^{(0)}, \theta^{(1)}, \ldots, \theta^{(n)}\}$ constitutes a set of representatives for the orbits of S_n on C_2^X. Clearly, the S_n-stabilizer of $\theta^{(k)}$ is isomorphic to $S_k \times S_{n-k}$. We recall that the irreducible representations of the symmetric group S_t are canonically parameterized by the partitions of t; see, for instance, [7, 11, 15, 37, 38, 61, 68, 73]. For $\lambda \vdash t$ (recall that this means that λ is a partition of t), we will denote by $(\rho^\lambda, S^\lambda)$ the irreducible representation of S_t canonically associated with λ (the representation space S^λ is called the *Young permutation module* or *Specht module* corresponding to λ). The irreducible representations of the group $S_k \times S_{n-k}$ are all of the form $\rho^\lambda \otimes \rho^\mu$, with $\lambda \vdash k$ and $\mu \vdash n - k$. We set

$$\rho_{[\lambda;\mu]} = \mathrm{Ind}_{C_2 \wr (S_k \times S_{n-k})}^{C_2 \wr S_n} [\, \widetilde{\chi}_{\theta^{(k)}} \otimes \overline{\rho^\lambda \otimes \rho^\mu} \,],$$

so that, applying Theorem 2.5.1, we have the following:

Theorem 2.5.3   *The set*

$$\{\rho_{[\lambda;\mu]} : \lambda \vdash k, \mu \vdash n - k \text{ and } 0 \le k \le n\}$$

constitutes a complete list of pairwise inequivalent irreducible representations of $(C_2 \wr S_n)$. $\square$

Basic results on the representation theory of the hyperoctahedral group may be found in [31].

2.6 Representation theory of groups of the form $F \wr S_n$

The representation theory of groups of the form $F \wr S_n$, with respect to the natural action of S_n on $X = \{1, 2, \ldots, n\}$, may be obtained just by specializing the theory that we have developed so far. We now show how to express the group $T_G(\sigma)$ (see (2.43)) in this case. Recall that an irreducible representation of the base group $F^X \cong F^n$ may be written in the form

$$\sigma = \bigotimes_{i=1}^{n} \sigma_i, \tag{2.47}$$

where $\sigma_1, \sigma_2, \ldots, \sigma_n \in \widehat{F}$. For $\tau \in \widehat{F}$ and σ as in (2.47), we denote by $n_\tau(\sigma)$ the number of $i \in \{1, 2, \ldots, n\}$ such that $\sigma_i \sim \tau$. Then, identifying S_0 with the trivial group, we have

$$T_G(\sigma) \cong \prod_{\tau \in \widehat{F}} S_{n_{\tau(\sigma)}},$$

where $\prod$ denotes the direct product of groups. Therefore (cf. Table 2.1 and Exercise 2.1.14)

$$I_{F \wr S_n}(\sigma) \cong F \wr \left(\prod_{\tau \in \widehat{F}} S_{n_{\tau(\sigma)}} \right) \cong \prod_{\tau \in \widehat{F}} \left(F \wr S_{n_{\tau(\sigma)}} \right). \tag{2.48}$$

We now recall some standard notation. If G is a finite group and $m \in \mathbb{N}$, we set $G^m = \underbrace{G \times G \times \cdots \times G}_{m \text{ times}}$ (if $m = 0$ we identify G^0 with the trivial group). If τ is a representation of G, we denote by $\tau^{\otimes m}$ the representation $\underbrace{\tau \otimes \tau \otimes \cdots \otimes \tau}_{m \text{ times}}$ (if $m = 0$, we identify $\tau^{\otimes 0}$ with the trivial representation). Then, another way to express (2.48) is to say that every $\sigma = \otimes_{i=1}^{n} \sigma_i \in \widehat{F^n}$ belongs to the same $(F \wr S_n)$-conjugacy class of the representation

$$\bigotimes_{\tau \in \widehat{F}} \tau^{\otimes n_\tau(\sigma)}. \tag{2.49}$$

Moreover, the representations of the form (2.49) constitute a complete sys tem of representatives for the $(F \wr S_n)$-conjugacy classes of $\widehat{F^n}$. That is, if we set

$$\Sigma(F, n) = \left\{ v = (n_\tau)_{\tau \in \widehat{F}} : n_\tau \geq 0, \sum_{\tau \in \widehat{F}} n_\tau = n \right\}$$

and, for $\nu = (n_\tau)_{\tau \in \widehat{F}} \in \Sigma(F, n)$,

$$\sigma_\nu = \bigotimes_{\tau \in \widehat{F}} \tau^{\otimes n_\tau}$$

then the set of such representatives is precisely given by

$$\{\sigma_\nu : \nu \in \Sigma(F, n)\}$$

(cf. (2.48)). For $\nu = (n_\tau)_{\tau \in \widehat{F}}$ we set

$$S_\nu = \prod_{\tau \in \widehat{F}} S_{n_\tau} \leq S_n.$$

Then, the isomorphism (cf. Exercise 2.1.14)

$$\prod_{\tau \in \widehat{F}} (F \wr S_{n_\tau}) \cong \left(F \wr \prod_{\tau \in \widehat{F}} S_{n_\tau} \right) = F \wr S_\nu$$

yields the equivalence

$$\bigotimes_{\tau \in \widehat{F}} \widetilde{\tau^{\otimes n_\tau}} \sim \widetilde{\bigotimes_{\tau \in \widehat{F}} \tau^{\otimes n_\tau}} \tag{2.50}$$

where $\widetilde{\tau^{\otimes n_\tau}}$ (resp. $\widetilde{\bigotimes_{\tau \in \widehat{F}} \tau^{\otimes n_\tau}}$) denotes the extension of $\tau^{\otimes n_\tau}$ (resp. $\bigotimes_{\tau \in \widehat{F}}$ $\tau^{\otimes n_\tau}$) from S_{n_τ} to $F \wr S_{n_\tau}$ (resp. from S_ν to $F \wr S_\nu$).

As a consequence, the representation theory of the groups of the form $F \wr S_n$ may be deduced from Theorem 2.4.4. Before stating the main result, we recall that, for $t \in \mathbb{N}$ and $\lambda \vdash t$, we denote by $(\rho^\lambda, S^\lambda)$ the irreducible S_t-representation associated with the partition λ; moreover, for $\nu = (n_\tau)_{\tau \in \widehat{F}} \in \Sigma(F, n)$ we set

$$\Lambda(\nu) = \{\lambda^{(\nu)} = (\lambda_\tau)_{\tau \in \widehat{F}} : \lambda_\tau \vdash n_\tau\}$$

and, for $\lambda = (\lambda_\tau)_{\tau \in \widehat{F}} \in \Lambda(\nu)$,

$$\rho_\lambda = \bigotimes_{\tau \in \widehat{F}} \rho^{\lambda_\tau}.$$

Using this notation we have:

Theorem 2.6.1 *The set*

$$\left\{ \mathrm{Ind}_{F \wr S_\nu}^{F \wr S_n} (\widetilde{\sigma_\nu} \otimes \overline{\rho_\lambda}) : \nu = (n_\tau)_{\tau \in \widehat{F}} \in \Sigma(F, n), \lambda \in \Lambda(\nu) \right\}$$

constitutes a complete list of pairwise inequivalent irreducible representations of $(F \wr S_n)$.

2.6.1 Representation theory of $S_m \wr S_n$

In the particular case when $F = S_m$, Theorem 2.6.1 gives (for $\mu \vdash m$ we simply write n_μ instead of n_{ρ^μ}):

Corollary 2.6.2 *Setting*

$$\sigma_\nu = \bigotimes_{\mu \vdash m} (\rho^\mu)^{\otimes n_\mu}, \quad \Lambda(\nu) = \{(\lambda_{n_\mu})_{\mu \vdash m} : \lambda_{n_\mu} \vdash n_\mu\} \quad and \quad \rho_\lambda = \bigotimes_{\mu \vdash m} \rho^{\lambda_\mu}$$

for every $\nu = (n_\mu)_{\mu \vdash m} \in \Sigma(S_m, n)$ and $\lambda \in \Lambda(\nu)$, we have that the set

$$\left\{\mathrm{Ind}_{S_m \wr S_\nu}^{S_m \wr S_n} (\widetilde{\sigma_\nu} \otimes \overline{\rho_\lambda}) : \nu = (n_\mu)_{\mu \vdash m} \in \Sigma(S_m, n), \lambda \in \Lambda(\nu)\right\} \qquad (2.51)$$

constitutes a complete list of pairwise inequivalent irreducible representations of $(S_m \wr S_n)$.

Remark 2.6.3 By virtue of the isomorphism (2.50) (and a similar isomorphism for the inflation) we can rewrite the above list of irreducible representations as

$$\left\{\mathrm{Ind}_{\prod_{\mu \vdash m} S_m \wr S_{n_\mu}}^{S_m \wr S_n} \left(\bigotimes_{\mu \vdash m} \widetilde{(\rho^\mu)^{\otimes n_\mu}} \otimes \overline{\rho^{\lambda_\mu}}\right) : \nu = (n_\mu)_{\mu \vdash m} \in \Sigma(S_m, n), \lambda \in \Lambda(\nu)\right\}$$

(cf. the list in [45, Section 4]).

Remark 2.6.4 In Example 2.3.13 we showed that the set of conjugacy classes of the group $S_m \wr S_n$ is parameterized by the set (2.42). From (2.51) we see that the same set, (2.42), parameterizes the set of all irreducible (pairwise inequivalent) representations of $S_m \wr S_n$. This is in accordance with the well-known fact that for any finite group F there is a bijective correspondence between the sets of conjugacy classes of F and of $\widehat{F}$, the dual of F.

3

Harmonic analysis on some homogeneous spaces of finite wreath products

The present chapter is an exposition of the results in the papers [9, 63, 64]. We have rearranged material in the original sources, adding more details and making everything consistent with the background developed in the preceding chapters. In the first section we examine the composition of two permutation representations (possibly with multiplicities, that is, not necessarily yielding Gelfand pairs) and present the corresponding decomposition into irreducible subrepresentations. We also give an explicit expression for the associated spherical matrix coefficients. In Section 3.2 we study the generalized Johnson scheme and describe a general construction of finite Gelfand pairs, introduced in [9], which is based on the action of the group $\mathrm{Aut}(T)$ of automorphisms of a finite rooted tree T on the space of all rooted subtrees of T. We then study the harmonic analysis of the exponentiation, following [64], and analyze in detail the lamplighter group by developing a harmonic analysis on the corresponding finite lamplighter spaces.

3.1 Harmonic analysis on the composition of two permutation representations

In this section we examine the composition of two permutation representations. We give the rule for decomposition into irreducible representations and the formulas for the related spherical matrix coefficients. For a more general treatment, namely for the harmonic analysis on the composition action of a crested product (a generalization of both the direct product and the wreath product), we refer to [64].

3.1.1 Decomposition into irreducible representations

Let G and F be finite groups and suppose that G (resp. F) acts transitively on a finite set X (resp. Y). If $\mathcal{G} \in L(X)$ and $g \in G$, we denote by $g\mathcal{G}$ the

g-image of $\mathcal{G}$, that is, the element in $L(X)$ defined by $(g\mathcal{G})(x) = \mathcal{G}(g^{-1}x)$ for all $x \in X$. Similarly, for $\mathcal{F} \in L(Y)$ and $f \in F$, we denote by $f\mathcal{F} \in L(Y)$ the f-image of $\mathcal{F}$. Suppose that

$$L(X) = \bigoplus_{i=0}^{n} a_i V_i \qquad (\text{resp. } L(Y) = \bigoplus_{j=0}^{m} b_j W_j) \qquad (3.1)$$

is the decomposition into irreducible G-representations (resp. F-represen-tations) of the corresponding permutation representation. Recall that, in our notation, $V_0, V_1, \ldots, V_n$ (resp. $W_0, W_1, \ldots, W_m$) are pairwise inequivalent irreducible representations and $a_0, a_1, \ldots, a_n$ (resp. $b_0, b_1, \ldots, b_m$) are their multiplicities in $L(X)$ (resp. in $L(Y)$); we also suppose that V_0 (resp. W_0) is the trivial representation and therefore, since the action is transitive, $a_0 = 1$ (resp. $b_0 = 1$). We fix $x_0 \in X$ and denote by $K = \{g \in G : gx_0 = x_0\}$ the stabilizer of x_0 in G.

For $\mathcal{G} \in L(X)$ and $\mathcal{F} \in L(Y)$ we identify the elementary tensor $\mathcal{G} \otimes \mathcal{F}$ with the element in $L(X \times Y)$ defined by

$$(x, y) \mapsto \mathcal{G}(x)\mathcal{F}(y) \qquad (3.2)$$

for all $x \in X$ and $y \in Y$. In this way, the set $\{\delta_x \otimes \delta_y : x \in X, y \in Y\}$ consti-tutes a basis for $L(X \times Y)$, and we have a natural isomorphism:

$$L(X \times Y) \cong L(X) \otimes L(Y). \qquad (3.3)$$

With the above notation, and considering $X \times Y$ as an $(F \wr G)$-permutation module with respect to the composition action (cf. (2.5)), we have the following.

Theorem 3.1.1 *The decomposition of $L(X \times Y)$ into irreducible $(F \wr G)$-subrepresentations is given by*

$$L(X \times Y) \cong \left[\bigoplus_{i=0}^{n} a_i(V_i \otimes W_0) \right] \bigoplus \left[\bigoplus_{j=1}^{m} b_j\big(L(X) \otimes W_j\big) \right]. \qquad (3.4)$$

Proof First, from (3.3) and (3.1) we get (3.4) as a vector space decomposition:

$$L(X \times Y) \simeq L(X) \otimes L(Y) = \left[\bigoplus_{i=0}^{n} a_i V_i \right] \otimes \left[\bigoplus_{j=0}^{m} b_j W_j \right]$$

$$= \left[\bigoplus_{i=0}^{n} a_i(V_i \otimes W_0) \right] \bigoplus \left[\bigoplus_{j=1}^{m} b_j\big(L(X) \otimes W_j\big) \right].$$

We have to show that every subspace on the right-hand side of (3.4) is invariant and irreducible.

Denote by λ the permutation representation of $F \wr G$ on $X \times Y$ associated with the composition action. Then the action of λ on tensor products can be expressed as follows. Let $\mathcal{G} \in L(X)$, $\mathcal{F} \in L(Y)$, $(x, y) \in X \times Y$ and $(f, g) \in F \wr G$. Recalling the formula for the inverse of (f, g) in Lemma 2.1.1, we have

$$
\begin{aligned}
[\lambda(f, g)(\mathcal{G} \otimes \mathcal{F})](x, y) &= (\mathcal{G} \otimes \mathcal{F})[(f, g)^{-1}(x, y)] \\
&= (\mathcal{G} \otimes \mathcal{F})[(g^{-1}f^{-1}, g^{-1})(x, y)] \\
&= (\mathcal{G} \otimes \mathcal{F})(g^{-1}x, f(x)^{-1}y) \\
&= (g\mathcal{G})(x)\,[f(x)\mathcal{F}]\,(y). \qquad (3.5)
\end{aligned}
$$

In general, the element in (3.5) is not an elementary tensor because $[f(x)\mathcal{F}]\,(y)$ depends on x. But there are two special instances where this is indeed the case. In fact, for $v \otimes \mathbf{1}_Y \in V_i \otimes W_0$, (3.5) gives

$$
\lambda(f, g)(v \otimes \mathbf{1}_Y) = gv \otimes \mathbf{1}_Y, \qquad (3.6)
$$

yielding the $(F \wr G)$-invariance of each space $V_i \otimes W_0$ (we denote by $\mathbf{1}_Y \in W_0$ the function on Y with constant value 1). Also, for $\delta_x \otimes w \in L(X) \otimes W_j$ and $(x', y') \in X \times Y$ we have, again by (3.5) and recalling that $g\delta_x = \delta_{gx}$,

$$
[\lambda(f, g)(\delta_x \otimes w)](x', y') = \delta_{gx}(x') \cdot [f(x')w](y').
$$

This implies that

$$
\lambda(f, g)(\delta_x \otimes w) = \delta_{gx} \otimes [f(gx)w], \qquad (3.7)
$$

showing that each space $L(X) \otimes W_j$ is invariant as well.

Let us now prove irreducibility. A representation of the form $V_i \otimes W_0$ is clearly irreducible, since V_i is G-irreducible and W_0 is trivial.

We could have obtained the same result as an application of Theorem 2.4.4. Indeed, taking σ as the trivial F^X-representation, that is, the tensor product $\bigotimes_{x \in X}(W_0)_x$ (recall that W_0 is the trivial representation of F), we have that the inertia group of σ is the whole of $F \wr G$, so that, by tensoring its extension to $F \wr G$ (which is the trivial representation of $F \wr G$) with V_i, we get exactly $V_i \otimes W_0$.

In order to show the irreducibility of the spaces $L(X) \otimes W_j$, we again use Theorem 2.4.4. Now, for $1 \leq j \leq m$ denote by σ_j the representation of F^X on the tensor product $\bigotimes_{x \in X} W_{\epsilon(x)}$, where $\epsilon(x_0) = j$ and $\epsilon(x) = 0$ for $x \neq x_0$. Let us check that the inertia group of σ_j is $F \wr K$. This follows from Lemma 2.4.2 upon observing that $T_G(\sigma_j) = \{g \in G : gx_0 = x_0\} = K$. Now, denoting by ι the trivial representation of $(F \wr K)/F^X \cong K$ and by $\tilde{\sigma}_j$ (resp. $\tilde{\iota}$) the extension of σ_j (resp. the inflation of ι) to $F \wr K$, let us show that the induced representation

$$\mathrm{Ind}_{F \wr K}^{F \wr G}(\bar{\iota} \otimes \tilde{\sigma}_j) \tag{3.8}$$

is equivalent to the representation of $F \wr G$ on $L(X) \otimes W_j$ (clearly $\bar{\iota} \otimes \tilde{\sigma}_j \sim \tilde{\sigma}_j$ and (3.8) is irreducible by Theorem 2.4.4). For each $x \in X$, choose $t_x \in G$ such that $t_x x_0 = x$. Then $\{t_x : x \in X\}$ is a system of representatives for the left cosets of K in G, and $T = \{(\mathbf{1}_F, t_x) : x \in X\}$ is a system of representatives for the left cosets of $F \wr K$ in $F \wr G$. From (3.7) we deduce that $\lambda(\mathbf{1}_F, t_x)\left[L(\{x_0\}) \otimes W_j\right] = \left[L(\{x\}) \otimes W_j\right]$, and therefore we have the direct sum decomposition

$$L(X) \otimes W_j = \bigoplus_{x \in X} \lambda(\mathbf{1}_F, t_x)\left[L(\{x_0\}) \otimes W_j\right]. \tag{3.9}$$

Another application of (3.7) yields, for $k \in K$, $f \in F^X$ and $w \in W_j$,

$$\lambda(f, k)(\delta_{x_0} \otimes w) = \delta_{x_0} \otimes [f(x_0)w] \tag{3.10}$$

and this shows that the representation of $F \wr K$ on $L(\{x_0\}) \otimes W_j$ is equivalent to $\bar{\iota} \otimes \tilde{\sigma}_j$ (see Lemma 2.4.3). By virtue of Proposition 1.1.9, from (3.9) and (3.10) we deduce that (3.8) is equivalent to $L(X) \otimes W_j$. It follows that $L(X) \otimes W_j$ is $(F \wr G)$-irreducible. $\qquad \square$

3.1.2 Spherical matrix coefficients

Keeping the same notation and assumptions as in the previous subsection, fix $y_0 \in Y$ and denote by $H \leq F$ its stabilizer. Let $X = \bigsqcup_{u=0}^{r} \Xi_u$ (resp. $Y = \bigsqcup_{v=0}^{t} \Lambda_v$) be the decomposition of X into K-orbits (resp. of Y into H-orbits), with $\Xi_0 = \{x_0\}$ (resp. $\Lambda_0 = \{y_0\}$).

Proposition 3.1.2

(i) *The stabilizer of (x_0, y_0) in $F \wr G$ is*

$$J = \{(f, k) \in F \wr G : k \in K, f(x_0) \in H\}.$$

(ii) *The decomposition of $X \times Y$ into J-orbits is*

$$X \times Y = \left[\bigsqcup_{v=0}^{t} (\Xi_0 \times \Lambda_v)\right] \bigsqcup \left[\bigsqcup_{u=1}^{s} (\Xi_u \times Y)\right].$$

Proof   (i) From (2.5) we have

$$(g, f)(x_0, y_0) = (g x_0, f(g x_0) y_0)$$

and therefore $(g, f)(x_0, y_0) = (x_0, y_0)$ if and only if $g \in K$ and $f(gx_0) \equiv f(x_0) \in H$.

(ii) If $(f, k) \in J$ and $(x_0, y) \in \Xi_0 \times \Lambda_v$ then

$$(f, k)(x_0, y) = (x_0, f(x_0)y) \in \Xi_0 \times \Lambda_v.$$

This show that $\Xi_0 \times \Lambda_v$ is J-invariant. Moreover, since $f(x_0)$, with $(f, g) \in J$, ranges among all elements of H, we deduce that J acts transitively on $\Xi_0 \times \Lambda_v$. This shows that the latter is a J-orbit. Analogously, if $(x, y) \in \Xi_u \times Y$, with $u \geq 1$, then

$$(f, k)(x, y) = (kx, f(kx)y) \in \Xi_u \times Y,$$

where the inclusion comes from the K-invariance of Ξ_u. Moreover, the action of J on $\Xi_u \times Y$ is transitive because on the one hand K acts transitively on Ξ_u and on the other hand $f(kx)$, with $(f, k) \in J$, ranges among all elements of F (which acts transitively on Y). $\qquad\square$

Remark 3.1.3   From Theorem 3.1.1 and Proposition 3.1.2(i) we deduce that if (G, K) and (F, H) are Gelfand pairs then $(F \wr G, J)$ is also a Gelfand pair.

Keeping in mind the decompositions in (3.1), suppose that

$$v_1^i, v_2^i, \ldots, v_{a_i}^i \qquad (\text{resp. } w_1^j, w_2^j, \ldots, w_{b_j}^j)$$

is an orthonormal basis for the subspace of K-invariant vectors in V_i, $i = 0, 1, \ldots, n$ (resp. for the subspace of H-invariant vectors in W_j, $j = 0, 1, \ldots, m$). Denoting by ρ^i (resp. σ^j) the representation of G on V_i (resp. of F on W_j), the corresponding spherical matrix coefficients (see Definition 1.2.14) are then given by

$$\phi_{\ell,r}^i(g) = \langle v_\ell^i, \rho^i(g)v_r^i \rangle_{V_i} \qquad (\text{resp. } \psi_{p,q}^j(f) = \langle w_p^j, \sigma^j(f)w_q^j \rangle_{W_j})$$

for $g \in G$, $i = 0, 1, \ldots, n$, and $\ell, r = 1, 2, \ldots, a_i$ (resp. for $f \in F$, $j = 0, 1, \ldots, m$, and $p, q = 1, 2, \ldots, b_j$).

Theorem 3.1.4

(i) *The elementary tensors*

$$v_\ell^i \otimes w_1^0, \qquad \ell = 1, 2, \ldots, a_i, \tag{3.11}$$

constitute an orthonormal basis for the J-invariant vectors in the irreducible representation $V_i \otimes W_0$, $i = 0, 1, \ldots, n$. Moreover, the associated spherical matrix coefficients are given by

$$\widetilde{\phi}_{\ell,r}^i(f, g) = \phi_{\ell,r}^i(g)$$

for $(f, g) \in F \wr G$, $i = 0, 1, \ldots, n$ and $\ell, r = 1, 2, \ldots, a_i$.

(ii) *The elementary tensors*

$$\delta_{x_0} \otimes w_p^j, \quad p = 1, 2, \dots, b_j, \tag{3.12}$$

constitute an orthonormal basis for the J-invariant vectors in the irreducible representation $L(X) \otimes W_j$, $j = 1, 2, \dots, m$. Moreover, the associated spherical matrix coefficients are given by

$$\widetilde{\psi}_{p,q}^j(f, g) = \mathbf{1}_K(g)\psi_{p,q}^j(f(x_0))$$

for $(f, g) \in F \wr G$, $j = 1, 2, \dots, m$ and $p, q = 1, 2, \dots, b_j$.

Proof   (i) It is clear that the elementary tensors in (3.11) constitute a set of a_i orthonormal vectors in $V_i \otimes W_0$. Therefore it suffices to show that these vectors are J-invariant. Now, applying (3.6) and noticing that $w_1^0 \equiv \frac{1}{|Y|}\mathbf{1}_Y$ we get, for all $(f, k) \in J$,

$$\lambda(f, k)(v_\ell^i \otimes w_1^0) = \rho^i(k)v_\ell^i \otimes w_1^0 = v_\ell^i \otimes w_1^0.$$

Moreover, for $(f, g) \in F \wr G$ we have

$$\begin{aligned}
\widetilde{\phi}_{\ell,r}^i(f, g) &= \left\langle v_\ell^i \otimes w_1^0, \lambda(f, g)(v_r^i \otimes w_1^0) \right\rangle_{V_i \otimes W_0} \\
&= \left\langle v_\ell^i \otimes w_1^0, \rho^i(g)v_r^i \otimes w_1^0 \right\rangle_{V_i \otimes W_0} \\
&= \left\langle v_\ell^i, \rho^i(g)v_r^i \right\rangle_{V_i} \left\langle w_1^0, w_1^0 \right\rangle_{W_0} \\
&= \phi_{\ell,r}^i(g).
\end{aligned}$$

(ii) Again, the elementary tensors in (3.12) constitute a set of b_j orthonormal vectors in $L(X) \otimes W_j$ and it suffices to show that these vectors are J-invariant. For all $(f, k) \in J$, applying (3.7) we get

$$\lambda(f, k)(\delta_{x_0} \otimes w_p^j) = \delta_{x_0} \otimes \left[\sigma^j(f(x_0))w_p^j \right] = \delta_{x_0} \otimes w_p^j,$$

because $f(x_0) \in H$ and w_p^j is H-invariant. Moreover,

$$\begin{aligned}
\widetilde{\psi}_{p,q}^j(f, g) &= \left\langle \delta_{x_0} \otimes w_p^j, \lambda(f, g)\left(\delta_{x_0} \otimes w_q^j\right) \right\rangle_{L(X) \otimes W_j} \\
&= \left\langle \delta_{x_0} \otimes w_p^j, \delta_{gx_0} \otimes \left[\sigma^j(f(gx_0))w_q^j \right] \right\rangle_{L(X) \otimes W_j} \\
&= \left\langle \delta_{x_0}, \delta_{gx_0} \right\rangle_{L(X)} \left\langle w_p^j, \left[\sigma^j(f(x_0))w_q^j \right] \right\rangle_{W_j} \\
&= \mathbf{1}_K(g)\psi_{p,q}^j(f(x_0)).
\end{aligned}$$

$\square$

Taking into account Proposition 3.1.2, we can describe the values of the spherical matrix coefficients on the orbits of J on $X \times Y$ (recall (1.38) and the isomorphism $L(X) \cong L(G)^K$), as follows.

Corollary 3.1.5

(i) *For $v = 0, 1, \ldots, t$ the value of $\widetilde{\phi}^i_{\ell,r}$ on $\Xi_0 \times \Lambda_v$ is equal to 1.*

(ii) *For $u = 1, 2, \ldots, s$ the value of $\widetilde{\phi}^i_{\ell,r}$ on $\Xi_u \times L(Y)$ is equal to the value of $\phi^i_{\ell,r}$ on Ξ_u.*

(iii) *For $v = 0, 1, \ldots, t$ the value of $\widetilde{\psi}^j_{p,q}$ on $\Xi_0 \times \Lambda_v$ is equal to the value of $\psi^i_{p,q}$ on Λ_v.*

(iv) *For $u = 1, 2, \ldots, s$ the value of $\widetilde{\psi}^j_{p,q}$ on $\Xi_u \times L(Y)$ is equal to 0.*

$\hfill\square$

3.2 The generalized Johnson scheme

3.2.1 The Johnson scheme

The Johnson scheme refers to the Gelfand pair $(S_n, S_{n-h} \times S_h)$ and is named after the American mathematician Selmer M. Johnson. The main sources are Delsarte's thesis [19] and the papers by Dunkl [25–27] and Stanton [70–72]; see also Chapter 6 of our monograph [11].

In what follows, n is a fixed positive integer and $0 \leq h \leq n$. We consider the action of S_n (resp. S_{n-h}, resp. S_h) on the set $\{1, 2, \ldots, n\}$ (resp. $\{1, 2, \ldots, n - h\}$, resp. $\{n - h + 1, n - h + 2, \ldots, n\}$) and denote by Ω_h the homogeneous space $S_n/(S_{n-h} \times S_h)$, which can be identified with the space of all h-subsets of $\{1, 2, \ldots, n\}$. The corresponding permutation module $L(\Omega_h)$ is denoted by $M^{n-h,h}$.

We introduce a metric δ on Ω_h by setting

$$\delta(A, B) = h - |A \cap B|$$

for all $A, B \in \Omega_h$. This is indeed a metric: its only nontrivial property is the triangular inequality, which follows immediately from

$$h = |B| \geq |A \cap B| + |B \cap C| - |A \cap B \cap C|$$
$$\geq |A \cap B| + |B \cap C| - |A \cap C|$$

for all $A, B, C \in \Omega_h$.

Proposition 3.2.1

(i) *$(S_n, S_{n-h} \times S_h)$ is a symmetric Gelfand pair.*

(ii) *The space $L(\Omega_h)$ decomposes into $\min\{n-h, h\}+1$ irreducible pairwise inequivalent S_n-representations.*

Proof   (i) By Example 1.2.32 it suffices to show that the action of the group S_n on the metric space (Ω_h, δ) is 2-point homogeneous. Let A, B, A', $B' \in \Omega_h$ and suppose that $\delta(A, B) = \delta(A', B')$. Then $|A \cap B| = |A' \cap B'|$, $|A \setminus B| = |A' \setminus B'|$ and $|B \setminus A| = |B' \setminus A'|$. Therefore we can find a permutation $\pi \in S_n$ such that $\pi(A \cap B) = A' \cap B'$, $\pi(A \setminus B) = A' \setminus B'$ and $\pi(B \setminus A) = B' \setminus A'$. It follows that $\pi(A) = A'$ and $\pi(B) = B'$, showing that the action is 2-point homogeneous.

(ii) Suppose first that $0 \leq h \leq n/2$. Then the range of δ is given by all integers between 0 and h and there are $h + 1$ ($S_{n-h} \times S_h$)-orbits on Ω_h; these are the spheres $\sigma_k = \{B \in \Omega_h : \delta(\overline{A}, B) = k\}$, $k = 0, 1, 2, \ldots, h$, where $\overline{A} = \{n-h+1, n-h+2, \ldots, n\}$ is the point in Ω_h stabilized by $S_{n-h} \times S_h$. By Corollary 1.2.35, $L(\Omega_h)$ decomposes into $h + 1$ irreducible pairwise inequivalent S_n-representations.

If $h > n/2$ then the argument is analogous: we just note that, in this case, the range of δ is given by all integers between 0 and $n - h$. $\square$

Since the spherical functions associated with the Gelfand pair $(S_n, S_{n-h} \times S_h)$ are bi-K-invariant (see Remark 1.2.16) and the characteristic functions of the $(S_{n-h} \times S_h)$-orbits constitute a basis of the space ${}^K L(G)^K$, we deduce from the proof of Proposition 3.2.1(ii) that the spherical functions are radial (they only depend on the distance from $\overline{A}$).

We define the operator $d : M^{n-h,h} \to M^{n-h+1,h-1}$ by setting

$$(d\gamma)(B) = \sum_{A \in \Omega_h : B \subseteq A} \gamma(A) \qquad (3.13)$$

for every $B \in \Omega_{h-1}$ and $\gamma \in M^{n-h,h}$. It is easy to see that the adjoint of d is the operator d^* defined by setting

$$(d^*\beta)(A) = \sum_{B \in \Omega_{h-1} : B \subseteq A} \beta(B) \qquad (3.14)$$

for every $A \in \Omega_h$ and $\gamma \in M^{n-h+1,h-1}$.

A proof of the following results may be found in [11, Theorems 6.1.6, 6.1.10 and 6.2.1] (we use the Pochhammer symbol notation $(a)_i = a(a+1)(a+2)\cdots(a+i-1)$ for $a \in \mathbb{C}$ and $i \in \mathbb{N}$).

Theorem 3.2.2

(i) *For $0 \leq k \leq n/2$, $M^{n-k,k} \cap \operatorname{Ker} d \equiv S^{n-k,k}$ is an irreducible representation of S_n and its dimension is equal to $\binom{n}{k} - \binom{n}{k-1}$.*

(ii) *If* $0 \leq k \leq \min\{n - h, h\}$ *then* $(d^*)^{h-k}$ *maps* $M^{n-k,k} \cap \operatorname{Ker} d$ *one to one into* $M^{n-h,h}$.

(iii) $M^{n-h,h} = \bigoplus_{k=0}^{\min\{n-h,h\}} (d^*)^{h-k}(M^{n-k,k} \cap \operatorname{Ker} d)$ *is the decomposition of* $M^{n-h,h}$ *into* S_n-*irreducible representations.*

(iv) *The spherical function* $\psi(n, h, k)$ *of* $(S_n, S_{n-h} \times S_h)$ *belonging to the subspace isomorphic to* $S^{n-k,k}$ *is given by*

$$\psi(n, h, k) = \sum_{u=0}^{\min\{n-h,h\}} \psi(n, h, k; u)\sigma_u, \tag{3.15}$$

where

$$\psi(n, h, k; u) = \frac{(-1)^k}{\binom{n-h}{k}} \sum_{i=\max\{0,u-h+k\}}^{\min\{u,k\}} \binom{h-u}{k-i}\binom{u}{i}$$
$$\times \frac{(n-h-k+1)_{k-i}}{(-h)_{k-i}}.$$

□

Notation 3.2.3 For $0 \leq u \leq v \leq n$ and $A \in \Omega_v$, $\Omega_u(A)$ will denote the space of all u-subsets of A. Also, we will denote by $M^{v-u,u}(A)$ the space $L(\Omega_u(A))$ seen as a module over the symmetric group S_v of all permutations of A (in this way, Ω_h coincides with $\Omega_h(\{1, 2, \ldots, n\})$).

3.2.2 The homogeneous space Θ_h

Let (F, H) be a finite Gelfand pair. We denote by $Y = F/H$ the correspondig homogeneous space and fix a point $y_0 \in Y$ stabilized by H. Let $Y = \bigsqcup_{i=0}^{m} \Lambda_i$ be the decomposition of Y into its H-orbits (with $\Lambda_0 = \{y_0\}$), $L(Y) = \bigoplus_{i=0}^{m} W_i$ the decomposition of $L(Y)$ into irreducible representations of F (with W_0 the trivial representation) and ϕ_i the spherical function in W_i, $i = 0, 1, \ldots, m$.

Let n be a positive integer and fix $0 \leq h \leq n$.

We will construct a natural homogeneous space for the wreath product $F \wr S_n$ using the actions of F on Y and of S_n on Ω_h.

Let Θ_h denote the set of all maps $\theta : A \to Y$, with A ranging in Ω_h. In other words,

$$\Theta_h = \coprod_{A \in \Omega_h} Y^A. \tag{3.16}$$

Bearing in mind the geometric interpretation of exponentiation (see Example 2.1.11 and Fig. 2.2), we interpret an element θ in Θ_h as a subtree of type $(h, 1)$ in the tree of $\{1, 2, \ldots, n\} \times Y$ (see Fig. 3.1).

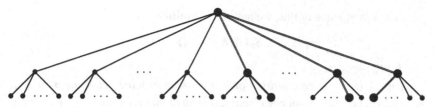

Fig. 3.1 An element $\theta \in \Theta_h$ coincides with a subtree of type $(h, 1)$ in the tree of $\{1, 2, \ldots, n\} \times Y$.

Let $A \in \Omega_h$ and $\theta \in Y^A \subset \Theta_h$. We then denote by $\text{dom}(\theta) = A$ the domain of definition of θ. The group $F \wr S_n$ acts transitively on Θ_h in a natural way: if $(f, \pi) \in F \wr S_n$ and $\theta \in \Theta_h$ then $(f, \pi)\theta$ is the element of Θ_h, with domain $\text{dom}((f, \pi)\theta) = \pi \text{dom}(\theta) \in \Omega_h$, defined by setting

$$[(f, \pi)\theta](j) = f(j)\theta(\pi^{-1}j) \in Y \qquad (3.17)$$

for every $j \in \pi \text{dom}(\theta)$.

Exercise 3.2.4

(i) Show that (3.17) defines an action of $F \wr S_n$ on Θ_h.
(ii) Show that this action is transitive.

Hint. Use the transitivity of S_n on Ω_h and the transitivity of F on Y.

Let $\overline{A} = \{n - h + 1, n - h + 2, \ldots, n\} \in \Omega_h$ denote the point stabilized by $S_{n-h} \times S_h$ and define $\theta_0 \in Y^{\overline{A}} \subseteq \Theta_h$ by setting $\theta_0(j) = y_0$ for every $j \in \overline{A}$.

Modulo the identification of $F \wr (S_{n-h} \times S_h)$ with $(F \wr S_{n-h}) \times (F \wr S_h)$ (cf. Exercise 2.1.14), we have the following.

Lemma 3.2.5 *The stabilizer of θ_0 in $F \wr S_n$ is equal to $(F \wr S_{n-h}) \times (H \wr S_h)$.*

Proof Let $(f, \pi) \in F \wr S_n$. Then $(f, \pi)\theta_0 = \theta_0$ if and only if $\pi(\overline{A}) = \overline{A}$ (so that necessarily $\pi \in S_{n-h} \times S_h$) and

$$f(j)\theta_0(\pi^{-1}(j)) = \theta_0(j) \qquad (3.18)$$

for all $j \in \text{dom}(\theta_0) = \{n - h + 1, n - h + 2, \ldots, n\}$. Since $\theta_0(j) = y_0$ for all $n - h + 1 \leq j \leq n$, (3.18) is equivalent to

$$f(j)y_0 = y_0, \qquad (3.19)$$

that is, $f(j) \in H$ for all $n - h + 1 \leq j \leq n$. In other words we have the decomposition $f = (f_{n-h}, f_h)$, where $f_{n-h} \in F^{\{1,2,\ldots,n-h\}}$ and $f_h \in H^{\{n-h+1,n-h+2,\ldots,n\}}$ denote the restrictions of f to $\{1, 2, \ldots, n - h\}$ and $\{n - h + 1, n - h + 2, \ldots, n\}$, respectively. $\square$

As a consequence of this, we have the identification

$$\Theta_h \cong (F \wr S_n)/[(F \wr S_{n-h}) \times (H \wr S_h)]$$

as $(F \wr S_n)$-spaces.

A *weak $(m + 1)$-composition* of h [69] is an ordered sequence $\mathbf{a} = (a_0, a_1, \ldots, a_m)$ of $m + 1$ nonnegative integers such that $a_0 + a_1 + \cdots + a_m = h$. The set of all weak $(m+1)$-compositions of h will be denoted by $C(h, m+1)$. It is easy to check that $|C(h, m + 1)| = \binom{m+h}{m}$. Indeed, the map

$$(a_0, a_1, \ldots, a_m) \mapsto \{a_0 + 1, a_0 + a_1 + 2, \ldots, a_0 + a_1 + \cdots + a_{m-1} + m\}$$

establishes a bijection between $C(h, m + 1)$ and the set of all m-subsets of $\{1, 2, \ldots, h + m\}$. For $\mathbf{a} = (a_0, a_1, \ldots, a_m) \in C(h, m + 1)$ we set $\ell(\mathbf{a}) = a_1 + a_2 + \cdots + a_m \equiv h - a_0$ and $\tilde{\mathbf{a}} = (a_1, a_2, \ldots, a_m)$. Clearly $\tilde{\mathbf{a}}$ is an element of $C(\ell(\mathbf{a}), m)$.

If $\mathbf{a} = (a_0, a_1, \ldots, a_m) \in C(h, m + 1)$ and $A \in \Omega_h$ then a *composition* of A of type $\mathbf{a}$ is an ordered sequence $\mathbf{A} = (A_0, A_1, \ldots, A_m)$ of subsets of A (necessarily disjoint but possibly empty) such that $A = \coprod_{i=0}^m A_i$ and $|A_i| = a_i$, $i = 0, 1, \ldots, m$. In other words $\mathbf{A}$ is an ordered partition of A. We denote by $\Omega_{\mathbf{a}}(A)$ the set of all compositions of A of type $\mathbf{a}$.

For the next definition, we recall that $\overline{A} = \{n - h + 1, n - h + 2, \ldots, n\} \in \Omega_h$ is the point stabilized by $S_{n-h} \times S_h$ and that $\Lambda_0 = \{y_0\}, \Lambda_1, \ldots, \Lambda_m \subset Y$ are the H-orbits on Y.

Definition 3.2.6   For every $\theta \in \Theta_h$, the *type* of θ is the sequence of $m + 2$ nonnegative integers

$$\mathrm{type}(\theta) = (t, b_0, b_1, \ldots, b_m),$$

where $t = t(\theta) = |\mathrm{dom}(\theta) \cap \overline{A}|$ and $b_i = b_i(\theta) = |\{j \in \mathrm{dom}(\theta) \cap \overline{A} : \theta(j) \in \Lambda_i\}|$, $i = 0, 1, \ldots, m$.

Lemma 3.2.7   *Two points $\theta_1, \theta_2 \in \Theta_h$ belong to the same orbit of $(F \wr S_{n-h}) \times (H \wr S_h)$ if and only if* $\mathrm{type}(\theta_1) = \mathrm{type}(\theta_2)$.

Proof   Let $\theta_1, \theta_2 \in \Theta_h$ and set $A_k = \mathrm{dom}(\theta_k) \in \Omega_h$ for $k = 1, 2$.

Suppose first that there exists $(f, \pi) \in (F \wr S_{n-h}) \times (H \wr S_h)$ such that $(f, \pi)(\theta_1) = \theta_2$. From $\pi = (\pi_{n-h}, \pi_h) \in S_{n-h} \times S_h$, $\pi(A_1) = A_2$ and $\pi(\overline{A}) = \overline{A}$ and the decompositions

$$A_k = (A_k \cap \overline{A}) \coprod (A_k \setminus \overline{A}),$$

$k = 1, 2$, we deduce

$$\pi_h(A_1 \cap \overline{A}) = A_2 \cap \overline{A} \quad \text{and} \quad \pi_{n-h}(A_1 \setminus \overline{A}) = A_2 \setminus \overline{A}$$

so that, in particular,

$$|A_1 \cap \overline{A}| = |A_2 \cap \overline{A}|.$$

This shows that $t(\theta_1) = t(\theta_2)$.

Let $0 \leq i \leq m$, $k = 1, 2$, and set $A_{k,i} = \{j \in A_k \cap \overline{A} : \theta_k(j) \in \Lambda_i\}$. Let us show that $|A_{1,i}| = |A_{2,i}|$. Let $j \in A_{2,i}$. From $(f, \pi)(\theta_1) = \theta_2$ and (3.17) we deduce that $f(j)\theta_1(\pi^{-1}(j)) = \theta_2(j)$. Since $j \in A_{2,i} \subset \overline{A}$, we have $f(j) \in H$ and therefore $\theta_1(\pi^{-1}(j)) \in \Lambda_i$ (since $\theta_2(j) \in \Lambda_i$). This shows that $\pi^{-1}(A_{2,i}) \subset A_{1,i}$. From the decomposition

$$A_k \cap \overline{A} = \coprod_{i=0}^{m} A_{k,i}$$

and using cardinalities, we deduce that indeed $\pi^{-1}(A_{2,i}) = A_{1,i}$. In particular $|A_{1,i}| = |A_{2,i}|$, that is, $b_i(\theta_1) = b_i(\theta_2)$. It follows that type$(\theta_1) = $ type(θ_2).

We leave the proof of the converse as an exercise. $\qquad\square$

Corollary 3.2.8 *The orbits of* $(H \wr S_h) \times (F \wr S_{n-h})$ *on* Θ_h *are parameterized by the set*

$$\{(t, \mathbf{b}) : \max\{0, 2h - n\} \leq t \leq h, \mathbf{b} \in C(t, m+1)\}$$

$$\equiv \coprod_{t=\max\{0, 2h-n\}}^{h} C(t, m+1).$$

Proof It suffices to note that if type$(\theta) = (t, b_0, b_1, \ldots, b_m)$ then $\sum_{i=0}^{m} b_i = t$ and $t = |\mathrm{dom}(\theta) \cap \overline{A}|$ is subject (only) to the condition $\max\{0, 2h - n\} \leq t \leq h$. $\qquad\square$

Remark 3.2.9 Suppose that $2h \leq n$. Then the map

$$(a_0, a_1, \ldots, a_m, a_{m+1}) \mapsto (a_0, a_1, \ldots, a_m)$$

establishes a bijection between $C(h, m+2)$ and $\coprod_{t=0}^{h} C(t, m+1)$. Analogously, the map

$$(a_0, a_1, \ldots, a_m, a_{m+1})$$
$$\mapsto (a_0, a_0 + a_1, \ldots, a_0 + a_1 + \cdots + a_{m-1}, a_0 + a_1 + \cdots + a_{m-1} + a_m)$$

is a bijection between $C(h, m+2)$ and the set $\{(i_1, i_2, \ldots, i_{m+1}) : 0 \leq i_1 \leq i_2 \leq \cdots \leq i_{m+1} \leq h\}$.

We end this subsection by defining two intertwining operators between the permutation representations on Θ_h and Θ_{h-1}. Suppose that $k < h$. Then for $\theta \in \Theta_h$ and $\xi \in \Theta_k$ we write $\xi \subseteq \theta$ when θ extends ξ, that is, when $\mathrm{dom}(\xi) \subseteq \mathrm{dom}(\theta)$ and $\theta|_{\mathrm{dom}(\xi)} = \xi$ (see Fig. 3.2).

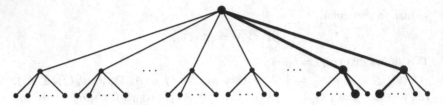

Fig. 3.2 An element $\xi \in \Theta_{h-1}$ such that $\xi \subseteq \theta$ (with θ as in Fig. 3.1).

Proposition 3.2.10 *Let $D : L(\Theta_h) \to L(\Theta_{h-1})$ be the linear operator defined by setting*

$$(DF)(\xi) = \sum_{\substack{\theta \in \Theta_h \\ \xi \subseteq \theta}} F(\theta) \qquad \text{for all } F \in L(\Theta_h) \text{ and } \xi \in \Theta_{h-1}.$$

Then D intertwines the permutation representations $L(\Theta_h)$ and $L(\Theta_{h-1})$. Moreover, its adjoint $D^* : L(\Theta_{h-1}) \to L(\Theta_h)$ is the linear operator defined by

$$(D^*G)(\theta) = \sum_{\substack{\xi \in \Theta_{h-1} \\ \xi \subseteq \theta}} G(\xi) \qquad \text{for all } G \in L(\Theta_{h-1}) \text{ and } \theta \in \Theta_h.$$

Proof Let $F \in L(\Theta_h)$, $\xi \in \Theta_{h-1}$ and $(f, \pi) \in F \wr S_n$. Then we have

$$[(f, \pi)DF](\xi) = [DF]\left((f, \pi)^{-1}\xi\right)$$

$$= \sum_{\substack{\theta \in \Theta_h: \\ (f,\pi)^{-1}\xi \subseteq \theta}} F(\theta)$$

$$= \sum_{\substack{\theta \in \Theta_h: \\ \xi \subseteq (f,\pi)\theta}} F(\theta)$$

$$= \sum_{\substack{\theta' \in \Theta_h: \\ \xi \subseteq \theta'}} F((f, \pi)^{-1}\theta') \qquad (\text{setting } \theta' = (f, \pi)\theta)$$

$$= \sum_{\substack{\theta' \in \Theta_h: \\ \xi \subseteq \theta'}} [(f, \pi)F](\theta')$$

$$= [D((f, \pi)F)](\xi).$$

This shows the first statement.

Moreover, if $\mathcal{F} \in \Theta_h$ and $\mathcal{G} \in \Theta_{h-1}$ then we have

$$\langle D\mathcal{F}, \mathcal{G} \rangle_{L(\Theta_{h-1})} = \sum_{\xi \in \Theta_{h-1}} D\mathcal{F}(\xi)\overline{\mathcal{G}(\xi)}$$

$$= \sum_{\xi \in \Theta_{h-1}} \left(\sum_{\substack{\theta \in \Theta_h: \\ \xi \subseteq \theta}} \mathcal{F}(\theta) \right) \overline{\mathcal{G}(\xi)}$$

$$= \sum_{\theta \in \Theta_h} \mathcal{F}(\theta) \sum_{\substack{\xi \in \Theta_{h-1}: \\ \xi \subseteq \theta}} \overline{\mathcal{G}(\xi)}$$

$$= \sum_{\theta \in \Theta_h} \mathcal{F}(\theta)\overline{D^*\mathcal{G}(\theta)}$$

$$= \langle \mathcal{F}, D^*\mathcal{G} \rangle_{L(\Theta_h)}.$$

This proves the second statement. □

3.2.3 Two special kinds of tensor product

Let $A \in \Omega_h$. We recall (cf. (3.3)) that there is a natural isomorphism between $L(Y)^{\otimes^h}$ and $L(Y^A)$. Explicitly, given $\mathcal{F}^j \in L(Y)$ for $j \in A$, the elementary tensor $\otimes_{j \in A}\mathcal{F}^j$ is interpreted as an element in $L(Y^A)$ by setting

$$\left(\otimes_{j \in A}\mathcal{F}^j \right)(\theta) = \prod_{j \in A} \mathcal{F}^j(\theta(j)) \qquad \text{for every } \theta \in Y^A. \tag{3.20}$$

Now we introduce another kind of tensor product. As above, we denote by $L(Y) = \bigoplus_{i=0}^m W_i$ the decomposition of the permutation representation $L(Y)$ into F-irreducibles, with W_0 the trivial representation. Let $\mathbf{a} = (a_0, a_1, \ldots, a_m) \in C(h, m+1)$, $B \in \Omega_{\ell(\mathbf{a})}$, $(A_1, A_2, \ldots, A_m) \in \Omega_{\tilde{\mathbf{a}}}(B)$, $\mathcal{F}^j \in W_i$ for $j \in A_i$, $i = 1, 2, \ldots, m$, and $\gamma \in M^{n-h,a_0}(\complement B)$ (see Notation 3.2.3); $\complement B$ is the complement of B. Then the elementary tensor $\gamma \otimes \left(\otimes_{j \in B}\mathcal{F}^j \right)$ is viewed as an element in

$$L\left(\coprod_{\substack{A \in \Omega_h: \\ B \subseteq A}} Y^A \right) \cong \bigoplus_{\substack{A \in \Omega_h: \\ B \subseteq A}} L(Y^A)$$

by setting

$$\left[\gamma \otimes \left(\otimes_{j \in B}\mathcal{F}^j \right) \right](\theta) = \gamma(\mathrm{dom}(\theta) \setminus B) \prod_{j \in B} \mathcal{F}^j(\theta(j)) \tag{3.21}$$

for every $\theta \in \Theta_h$ satisfying the condition $\mathrm{dom}(\theta) \supseteq B$.

Clearly, a tensor product as in (3.21) may be expressed by means of elementary tensors of the first kind:

$$\gamma \otimes \left(\otimes_{j\in B}\mathcal{F}^j\right) = \sum_{A_0'\in\Omega_{a_0}(\complement B)} \gamma(A_0')\left[\left(\otimes_{j\in A_0'}\phi_0\right) \otimes \left(\otimes_{j\in B}\mathcal{F}^j\right)\right], \quad (3.22)$$

where $\left(\otimes_{j\in A_0'}\phi_0\right)$ is the characteristic function of $Y^{A_0'}$ (and each summand is viewed as a simple tensor of the first kind acting on $Y^{A'}$ with $A' = A_0' \coprod B$).

In the following we describe the action of the group $F \wr S_n$ on such tensor products.

Lemma 3.2.11 *The action of $F \wr S_n$ on the tensor products introduced above is given by*

$$(f, \pi)\left(\otimes_{j\in A}\mathcal{F}^j\right) = \otimes_{t\in\pi A} f(t)\mathcal{F}^{\pi^{-1}t} \quad (3.23)$$

and

$$(f, \pi)\left[\gamma \otimes \left(\otimes_{j\in B}\mathcal{F}^j\right)\right] = (\pi\gamma) \otimes \left(\otimes_{t\in\pi B} f(t)\mathcal{F}^{\pi^{-1}t}\right) \quad (3.24)$$

for all $(f, \pi) \in F \wr S_n$.

Proof To show (3.23) let $\theta \in Y^{\pi A}$. Then

$$\left[(f, \pi)\left(\otimes_{j\in A}\mathcal{F}^j\right)\right](\theta) = \left(\otimes_{j\in A}\mathcal{F}^j\right)[(f, \pi)^{-1}\theta]$$

$$= \prod_{j\in A} \mathcal{F}^j\{[(f, \pi)^{-1}\theta](j)\}$$

$$= \prod_{j\in A} \mathcal{F}^j[f(\pi j)^{-1}\theta(\pi j)]$$

$$= \prod_{t\in\pi A} [f(t)\mathcal{F}^{\pi^{-1}t}](\theta(t))$$

$$= \left[\otimes_{t\in\pi A} f(t)\mathcal{F}^{\pi^{-1}t}\right](\theta).$$

Finally, (3.24) follows immediately from the decomposition (3.22) and from (3.23).   □

We now present two formulas that express the action of the operators D and D^* (see Proposition 3.2.10) on a tensor product of the second kind in terms of the action of the operators d (see (3.13)) and d^* (see (3.14)). Here we consider an element in

$$\bigoplus_{A \supseteq B : |A| = h} L(Y^A) \qquad \left(\text{resp. } \bigoplus_{A' \supseteq B : |A'| = h-1} L(Y^{A'})\right)$$

as an element in $L(\Theta_h)$ (resp. $L(\Theta_{h-1})$) which does not depend on the $Y^{A''}$ satisfying $|A''| = h$ (resp. $|A''| = h - 1$) and $A'' \not\supseteq B$.

Lemma 3.2.12 *We have*

$$D\left[\gamma \otimes \left(\otimes_{j \in B} \mathcal{F}^j\right)\right] = |Y|\left[(d\gamma) \otimes \left(\otimes_{j \in B} \mathcal{F}^j\right)\right]$$

and

$$D^*\left[\gamma \otimes \left(\otimes_{j \in B} \mathcal{F}^j\right)\right] = (d^*\gamma) \otimes \left(\otimes_{j \in B} \mathcal{F}^j\right).$$

Proof Since $\left[\gamma \otimes \left(\otimes_{j \in B} \mathcal{F}^j\right)\right](\theta)$ is defined for those θ's satisfying the condition $\text{dom}(\theta) \supseteq B$ then $\left(D\left[\gamma \otimes \left(\otimes_{j \in B} \mathcal{F}^j\right)\right]\right)(\xi)$ is defined for those $\xi \in \Theta_{h-1}$ satisfying the condition $|B \setminus \text{dom}(\xi)| \leq 1$, that is, for those ξ's for which there exists $\theta \in \Theta_h$ such that $\text{dom}(\theta) \supseteq B$ and $\xi \subseteq \theta$. On the one hand, if $|B \setminus \text{dom}(\xi)| = 0$, that is, $\text{dom}(\xi) \supseteq B$, then

$$\left(D\left[\gamma \otimes \left(\otimes_{j \in B} \mathcal{F}^j\right)\right]\right)(\xi)$$

$$= \sum_{\substack{\theta \in \Theta_h : \theta \supseteq \xi \\ \text{dom}\theta \supseteq B}} \left[\gamma \otimes \left(\otimes_{j \in B} \mathcal{F}^j\right)\right](\theta)$$

$$= \sum_{\theta \in \Theta_h : \theta \supseteq \xi} \gamma(\text{dom}(\theta) \setminus B) \prod_{j \in B} \mathcal{F}^j(\theta(j))$$

$$= \sum_{v \in \complement \text{dom}\xi} \sum_{y \in Y} \gamma[(\text{dom}(\xi) \sqcup \{v\}) \setminus B] \prod_{j \in B} \mathcal{F}^j(\theta(j))$$

$$= |Y|(d\gamma)(\text{dom}(\xi) \setminus B) \prod_{j \in B} \mathcal{F}^j(\xi(j))$$

$$= |Y|\left[(d\gamma) \otimes \left(\otimes_{j \in B} \mathcal{F}^j\right)\right](\xi)$$

(where the sum $\sum_{y \in Y}$, or equivalently the factor $|Y|$, comes from the fact that we have $|Y|$ different possible extensions of a function $\xi \Theta_{h-1}$ to a function $\theta \in \Theta_h$ with $\text{dom}(\theta) \supset \text{dom}(\xi)$). On the other hand, if $|B \setminus \text{dom}(\xi)| = 1$ and u is the unique element in $B \setminus \text{dom}(\xi)$ then

$$\left(D\left[\gamma \otimes \left(\otimes_{j \in B} \mathcal{F}^j\right)\right]\right)(\xi) = \gamma[(\text{dom}(\xi) \coprod \{u\}) \setminus B] \left(\sum_{y \in Y} \mathcal{F}^u(y)\right)$$

$$\times \prod_{j \in B \setminus \{u\}} \mathcal{F}^j(\xi(j)) = 0,$$

since $\mathcal{F}^u \notin W_0$. In particular, if $a_0 = 0$ then $D\left[\gamma \otimes \left(\otimes_{j \in B}\mathcal{F}^j\right)\right] = 0$.
The proof of (ii) is left to the reader.   □

3.2.4 The decomposition of $L(\Theta_h)$ into irreducible representations

We recall that $L(Y) = \oplus_{i=0}^{m} W_i$ is the decomposition of $L(Y)$ into F-irreducible representations.

Definition 3.2.13 Let $A \in \Omega_h$, $\mathbf{a} = (a_0, a_1, \ldots, a_m) \in C(h, m+1)$ and $\mathbf{A} = (A_0, A_1, \ldots, A_m) \in \Omega_{\mathbf{a}}(A)$. Then

(i) the space $W_{\mathbf{a}}(\mathbf{A})$ is the subspace of $L(Y^A)$ spanned by all the tensor products $\otimes_{j \in A}\mathcal{F}^j$ such that $\mathcal{F}^j \in W_i$ for every $j \in A_i$, $i = 0, 1, \ldots, m$;
(ii) we then set

$$W_{h,\mathbf{a}} = \bigoplus_{A \in \Omega_h} \bigoplus_{\mathbf{A} \in \Omega_{\mathbf{a}}(A)} W_{\mathbf{a}}(\mathbf{A}).$$

It is clear that $W_{h,\mathbf{a}}$ coincides with the subspace of $L(\Theta_h)$ spanned by all the tensor products $\gamma \otimes \left(\otimes_{j \in B}\mathcal{F}^j\right)$ where $B \in \Omega_{\ell(\mathbf{a})}$, $\gamma \in M^{n-h,a_0}(\complement B)$, and such that there exists $(A_1, A_2, \ldots, A_m) \in \Omega_{\widetilde{\mathbf{a}}}(B)$ satisfying $\mathcal{F}^j \in W_i$ for all $j \in A_i$ and $i = 1, 2, \ldots, m$.

Lemma 3.2.11 ensures that each $W_{h,\mathbf{a}}$ is an $(F \wr S_n)$-invariant subspace of $L(\Theta_h)$.

Lemma 3.2.14 For $\mathbf{a} \in C(h, m+1)$ we have

$$W_{h,\mathbf{a}} = \mathrm{Ind}_{F \wr S_{n-h} \times F \wr S_{a_0} \times F \wr S_{a_1} \times \cdots \times F \wr S_{a_m}}^{F \wr S_n} \left(I_{F \wr S_{n-h}} \otimes W_0^{\otimes a_0} \otimes W_1^{\otimes a_1} \right.$$
$$\left. \otimes \cdots \otimes W_m^{\otimes a_m} \right),$$

where $I_{F \wr S_{n-h}}$ denotes the identity representation of $F \wr S_{n-h}$.

Proof From Proposition 2.1.3 we deduce that

$$(F \wr S_n)/((F \wr S_{n-h}) \times (F \wr S_{a_0}) \times (F \wr S_{a_1}) \times \cdots \times (F \wr S_{a_m}))$$
$$\equiv S_n/(S_{n-h} \times S_{a_0} \times S_{a_1} \times \cdots \times S_{a_m}) \equiv \bigsqcup_{A \in \Omega_h} \bigsqcup_{\mathbf{A} \in \Omega_{\mathbf{a}}(A)} \mathbf{A}.$$

Moreover, if $\overline{\mathbf{A}} = (A_0, A_1, \ldots, A_m) \in \Omega_{\mathbf{a}}(\overline{A})$ is the composition stabilized by $S_{n-h} \times S_{a_0} \times S_{a_1} \times \cdots \times S_{a_m}$ (so that S_{a_i} is the symmetric group on $A_i = \{n - h + a_0 + a_1 + \cdots + a_{i-1} + 1, \ldots, n - h + a_0 + a_1 + \cdots + a_i\}$, $i = 0, 1, \ldots, m$) then $W_{\mathbf{a}}(\overline{\mathbf{A}})$, as a representation of $(F \wr S_{n-h}) \times (F \wr S_{a_0}) \times (F \wr S_{a_1}) \times \cdots \times (F \wr S_{a_m})$, is equivalent to $I_{F \wr S_{n-h}} \otimes W_0^{\otimes a_0} \otimes W_1^{\otimes a_1} \otimes \cdots \otimes W_m^{\otimes a_m}$. With these considerations, the lemma follows from the definition of $W_{h,\mathbf{a}}$.   □

From the definition of an induced operator (see (1.21)) and from Lemmas 3.2.12 and 3.2.14, we immediately get the following description of the operator D and its adjoint D^*.

Corollary 3.2.15

(i) $D = |Y| \operatorname{Ind}_{F \wr S_{n-\ell(\mathbf{a})} \times F \wr S_{a_1} \times \cdots \times F \wr S_{a_m}}^{F \wr S_n} d \otimes I \otimes \cdots \otimes I;$

(ii) $D^* = \operatorname{Ind}_{F \wr S_{n-\ell(\mathbf{a})} \times F \wr S_{a_1} \times \cdots \times F \wr S_{a_m}}^{F \wr S_n} d^* \otimes I \otimes \cdots \otimes I.$

$\square$

Recalling the notation of Theorem 3.2.2(i), we introduce the following representation.

Definition 3.2.16 *For $0 \leq k \leq (n - \ell(\mathbf{a}))/2$ we set*

$$W_{h,\mathbf{a},k} = \operatorname{Ind}_{F \wr S_{n-\ell(\mathbf{a})} \times F \wr S_{a_1} \times \cdots \times F \wr S_{a_m}}^{F \wr S_n} S^{n-\ell(\mathbf{a})-k,k} \otimes W_1^{\otimes a_1} \otimes \cdots \otimes W_m^{\otimes a_m}.$$

Recalling the definition of a *multinomial coefficient*, namely

$$\binom{n}{k_0, k_1, \ldots, k_m} = \frac{n!}{k_0! k_1! \cdots k_m!},$$

where $n, k_0, k_1, \ldots, k_m \in \mathbb{N}$ satisfy $n = \sum_{i=0}^m k_i$, it is clear that

$$\dim W_{h,\mathbf{a},k} = \binom{n}{n - \ell(\mathbf{a}), a_1, \ldots, a_m} \left[\binom{n - \ell(\mathbf{a}) - k}{k} - \binom{n - \ell(\mathbf{a}) - k}{k - 1} \right]$$
$$\times (\dim W_1)^{a_1} (\dim W_2)^{a_2} \cdots (\dim W_m)^{a_m}.$$

Lemma 3.2.17 *We have the orthogonal direct sum decomposition*

$$W_{h,\mathbf{a}} = \bigoplus_{k=0}^{\min\{n-h, h-\ell(\mathbf{a})\}} W_{h,\mathbf{a},k}.$$

Proof Using the transitivity of induction (see Proposition 1.1.10) we get

$$\operatorname{Ind}_{F \wr S_{n-h} \times F \wr S_{a_0} \times F \wr S_{a_1} \times \cdots \times F \wr S_{a_m}}^{F \wr S_n}$$

$$= \operatorname{Ind}_{F \wr S_{n-h+a_0} \times F \wr S_{a_1} \times \cdots \times F \wr S_{a_m}}^{F \wr S_n} \operatorname{Ind}_{F \wr S_{n-h} \times F \wr S_{a_0} \times F \wr S_{a_1} \times \cdots \times F \wr S_{a_m}}^{F \wr S_{n-h+a_0} \times F \wr S_{a_1} \times \cdots \times F \wr S_{a_m}}$$

and, since

$$\operatorname{Ind}_{F \wr S_{n-h} \times F \wr S_{a_0} \times F \wr S_{a_1} \times \cdots \times F \wr S_{a_m}}^{F \wr S_{n-h+a_0} \times F \wr S_{a_1} \times \cdots \times F \wr S_{a_m}} \left(I_{F \wr S_{n-h}} \otimes W_0^{\otimes a_0} \otimes W_1^{\otimes a_1} \otimes \otimes W_m^{\otimes a_m} \right)$$

$$- M^{n \ h, a_0} \otimes W_1^{\otimes a_1} \otimes \cdots \otimes W_m^{\otimes a_m}$$

($I_{F \wr S_{n-h}} \otimes W_0^{\otimes a_0}$ is the trivial representation of $(F \wr S_{n-h}) \times (F \wr S_{a_0})$), the lemma follows from the decomposition $M^{n-h,a_0} = \bigoplus_{k=0}^{\min\{n-h,a_0\}} S^{n-h+a_0-k,k}$ from Theorem 3.2.2(iii).

$\square$

In the following, we show how to construct the representations $W_{h,\mathbf{a},k}$ using the operators D and D^*. We set $\mathbf{a}^{(-k)} = (a_0 - k, a_1, \ldots, a_m)$.

Corollary 3.2.18

(i) $W_{k+\ell(\mathbf{a}),\mathbf{a},k} = \operatorname{Ker} D \cap W_{k+\ell(\mathbf{a}),\mathbf{a}}$;

(ii) *If* $0 \leq k \leq \min\{n - h, h - \ell(\mathbf{a})\}$ *then* $(D^*)^{h-k-\ell(\mathbf{a})}$ *is an isomorphism of* $W_{k+\ell(\mathbf{a}),\mathbf{a}^{(-k)},k}$ *onto* $W_{h,\mathbf{a},k}$.

Proof This is an immediate consequence of Proposition 1.1.18, Theorem 3.2.2, Corollary 3.2.15 and Lemma 3.2.17.   □

We are now in a position to present the main result of this section and to introduce the *generalized Johnson scheme*.

Theorem 3.2.19

(i) *The set* $\{W_{h,\mathbf{a},k} : \mathbf{a} \in C(h, m + 1), \ 0 \leq k \leq \min\{n - h, h - \ell(\mathbf{a})\}\}$ *consists of the pairwise inequivalent irreducible representations of* $F \wr S_n$.

(ii) $(F \wr S_n, (H \wr S_h) \times (F \wr S_{n-h}))$ *is a Gelfand pair (the* generalized Johnson *scheme).*

(iii) *The decomposition of* $L(\Theta_h)$ *into irreducible* $(F \wr S_n)$*-representations is given by*

$$L(\Theta_h) = \bigoplus_{\mathbf{a} \in C(h,m+1)} \bigoplus_{k=0}^{\min\{n-h,h-\ell(\mathbf{a})\}} W_{h,\mathbf{a},k}. \tag{3.25}$$

Proof First note that from (3.16) we can immediately deduce the following decomposition:

$$L(\Theta_h) = \bigoplus_{A \in \Omega_h} L(Y^A). \tag{3.26}$$

Moreover, from the decomposition $L(Y) = \bigoplus_{i=0}^{m} W_i$ (into irreducible F-representations) and from the definition of $W_{\mathbf{a}}(\mathbf{A})$ it follows that

$$L(Y^A) \cong L(Y)^{\otimes^h} = \bigoplus_{l_1=0}^{m} \bigoplus_{l_2=0}^{m} \cdots \bigoplus_{l_h=0}^{m} W_{l_1} \otimes W_{l_2} \cdots \otimes W_{l_h}$$

$$= \bigoplus_{\mathbf{a} \in C(h,m+1)} \bigoplus_{\mathbf{A} \in \Omega_{\mathbf{a}}(A)} W_{\mathbf{a}}(\mathbf{A}). \tag{3.27}$$

From (3.26), (3.27) and the definition of $W_{h,\mathbf{a}}$ we deduce that $L(\Theta_h) = \bigoplus_{\mathbf{a}\in C(h,m+1)} W_{h,\mathbf{a}}$. Therefore from Lemma 3.2.17 it follows that (3.25) is an orthogonal decomposition of $L(\Theta_h)$ into $(F \wr S_n)$-invariant subspaces. Now, the map

$$T(t, b_0, b_1, \ldots, b_m)$$

$$= \begin{cases} (t + n - 2h, b_0 + h - t, b_1, \ldots, b_m) & \text{if } n - h < h - \ell(\mathbf{b}) \\ (b_0, b_0 + h - t, b_1, \ldots, b_m) & \text{if } n - h \geq h - \ell(\mathbf{b}) \end{cases}$$

is a bijection between the set in Corollary 3.2.8, which parameterizes the $((H \wr S_n) \times (F \wr S_{n-h}))$-orbits on Θ_h, and the set

$$\{(k, a_0, a_1, \ldots, a_m) : 0 \leq k \leq \min\{n - h, h - \ell(\mathbf{a})\}, \mathbf{a} \in C(h, m + 1)\},$$

which parameterizes the $(F \wr S_n)$-subrepresentations of $L(\Theta_h)$ in (3.25). The corresponding inverse map is given by

$$T^{-1}(k, a_0, a_1, \cdots, a_m)$$

$$= \begin{cases} (k - n + 2h, a_0 + k - n + h, a_1, \ldots, a_m) & \text{if } n - h < h - \ell(\mathbf{a}) \\ (k + h - a_0, k, a_1, \ldots, a_m) & \text{if } n - h \geq h - \ell(\mathbf{a}). \end{cases}$$

Therefore the three statements follow from the criterion for Gelfand pairs given in Theorem 1.2.36. $\square$

Remark 3.2.20 The result in Theorem 3.2.19(i) may be also obtained from the general representation theory of the wreath product $F \wr S_n$ (see Section 2.6). Indeed, $V = W_1^{\otimes a_1} \otimes \cdots \otimes W_m^{\otimes a_m}$ is an irreducible representation of the base group $F^{\times^n}$, the inertia group of V is $F \wr (S_{n-h+a_0} \times S_{a_1} \otimes \cdots \otimes S_{a_m})$, $S^{n-h+a_0-k,k}$ is an irreducible representation of $S_{n-h+a_0} \times S_{a_1} \times \cdots \times S_{a_m}$ (trivial on $S_{a_1} \times \cdots \times S_{a_m}$) and $W_{h,\mathbf{a},k}$ is obtained by the induction of $S^{n-h+a_0-k,k} \otimes V$ from the inertia group to $F \wr S_n$.

3.2.5 The spherical functions

Let $\mathbf{a} = (a_0, a_1, \ldots, a_m) \in C(h, m + 1)$. For $0 \leq u \leq \min\{n - h, h - \ell(\mathbf{a})\}$ we define the function

$$\Phi(h, \mathbf{a}, u)$$

$$= \sum_{\substack{(A_1, A_2, \ldots, A_m) \\ \in \Omega_{\bar{\mathbf{a}}}(\bar{A})}} \sum_{\substack{A_0 \in \Omega_{a_0}(\complement(A_1 \cup \cdots \cup A_m)): \\ |A_0 \setminus \bar{A}| = u}} \left[(\otimes_{j \in A_0} \phi_0) \otimes (\otimes_{j \in A_1} \phi_1) \otimes \cdots \otimes (\otimes_{j \in A_m} \phi_m) \right],$$

$$(3.28)$$

where $\left(\otimes_{j \in A_i} \phi_i\right)$ indicates the tensor product of a_i copies of the spherical function ϕ_i. From Lemma 3.2.11 we deduce that each $\Phi(h, \mathbf{a}, u)$ is $((H \wr S_h) \times (F \wr S_{n-h}))$-invariant. Moreover, the set

$$\{\Phi(h, \mathbf{a}, u) : 0 \le u \le \min\{n - h, h - \ell(\mathbf{a})\}\} \tag{3.29}$$

constitutes an orthogonal basis for the space of all $((H \wr S_h) \times (F \wr S_{n-h}))$-invariant functions in the module $W_{h,\mathbf{a}}$. Indeed, we have

$$\Phi(h, \mathbf{a}, u) \in \bigoplus_{\substack{B_1 \in \Omega_{h-u}(\overline{A}) \\ B_2 \in \Omega_u(\overline{CA})}} L(Y^{B_1 \sqcup B_2})$$

and the summands on the right-hand side are orthogonal for different values of u. The spherical functions can be expressed as linear combinations of the $\Phi(h, \mathbf{a}, u)$'s. We will use the notation of (3.15).

Theorem 3.2.21 *The spherical function* $\Psi(n, h, \mathbf{a}, k)$ *in* $W_{h,\mathbf{a},k}$ *is given by*

$$\Psi(n, h, \mathbf{a}, k)$$
$$= \frac{1}{\binom{h}{a_0, a_1, \ldots, a_m}} \sum_{u=0}^{\min\{n-h, h-\ell(\mathbf{a})\}} \psi(n - \ell(\mathbf{a}), h - \ell(\mathbf{a}), k; u)\Phi(h, \mathbf{a}, u).$$

Proof The function $\Psi(n, h, \mathbf{a}, k)$ is $((H \wr S_h) \times (F \wr S_{n-h}))$-invariant because it is a linear combination of invariant functions. Moreover, its value on θ_0, the point stabilized by $(H \wr S_h) \times (F \wr S_{n-h})$, is equal to 1. From (3.28) it follows that

$$\sum_{u=0}^{\min\{n-h, a_0\}} \psi(n - h + a_0, a_0, k; u)\Phi(h, \mathbf{a}, u)$$

$$= \sum_{\substack{(A_1, A_2, \ldots, A_m) \\ \in \Omega_{\tilde{\mathbf{a}}}(\overline{A})}} \sum_{u=0}^{\min\{n-h, a_0\}} \psi(n - h + a_0, a_0, k; u)$$

$$\times \sum_{\substack{A_0 \in \Omega_{a_0}(\mathbb{C}(A_1 \cup \cdots \cup A_m)): \\ |A_0 \setminus \overline{A}| = u}} \left[(\otimes_{j \in A_0} \phi_0) \otimes (\otimes_{j \in A_1} \phi_1) \otimes \cdots \otimes (\otimes_{j \in A_m} \phi_m)\right]$$

$$= \sum_{\substack{(A_1, A_2, \ldots, A_m) \\ \in \Omega_{\tilde{\mathbf{a}}}(\overline{A})}} \left[\psi(n - h + a_0, a_0, k) \otimes (\otimes_{j \in A_1} \phi_1) \otimes \cdots \otimes (\otimes_{j \in A_m} \phi_m)\right],$$

since

$$\sum_{u=0}^{\min\{n-h,a_0\}} \psi(n-h+a_0, a_0, k; u) \sum_{\substack{A_0 \in \Omega_{a_0}(\complement(A_1 \cup \cdots \cup A_m)): \\ |A_0 \setminus \overline{A}| = u}} \left(\otimes_{j \in A_0} \phi_0 \right)$$

coincides with the spherical function of the Gelfand pair $(S_{n-h+a_0}, S_{n-h} \times S_{a_0})$ belonging to the irreducible representation $S^{n-h+a_0-k,k}$ (here S_{n-h+a_0} is the symmetric group on $\complement(A_1 \cup \cdots \cup A_m)$ and $S_{n-h} \times S_{a_0}$ is the stabilizer of $\overline{A} \setminus (A_1 \cup A_2 \cup \cdots \cup A_m)$). Hence $\psi(n-h+a_0, a_0, k) \otimes \left(\otimes_{j \in A_1} \phi_1 \right) \otimes \cdots \otimes \left(\otimes_{j \in A_m} \phi_m \right)$ belongs to $S^{n-h+a_0-k,k} \otimes W_1^{\otimes a_1} \otimes \cdots \otimes W_m^{\otimes a_m}$ and, bearing in mind Definition 3.2.16, the theorem follows. $\square$

In the remaining part of this section, we first give the value of the spherical functions on a fixed $((H \wr S_h) \times (F \wr S_{n-h}))$-orbit. Then we specify the above analysis to the particular case when (F, H) is the Gelfand pair of the ultrametric space (see Section 2.1.5).

We denote by $\phi_i(j)$ the value of the spherical function ϕ_i on the orbit Λ_j. The value of $\Phi(h, \mathbf{a}, u)$ on a map $\theta \in \Theta_h$ with type$(\theta) = (t, \mathbf{b})$ is equal to 0 if $t = |\text{dom}(\theta) \cap \overline{A}| \neq h - u$, while if $t = h - u$ then the value is equal to

$$\Phi(h, \mathbf{a}, u; \mathbf{b}) = \sum_{(\alpha_{ij})} \prod_{j=0}^{m} \binom{b_j}{\alpha_{0j}, \alpha_{1j}, \ldots, \alpha_{mj}} \prod_{i=0}^{m} [\phi_i(j)]^{\alpha_{ij}}, \qquad (3.30)$$

where the sum is over all nonnegative integer-valued matrices

$$(\alpha_{ij})_{\substack{i=0,1,\ldots,m \\ j=0,1,\ldots,m}}$$

such that $\sum_{i=0}^{m} \alpha_{ij} = b_j$, $j = 0, 1, \ldots, m$, $\sum_{j=0}^{m} \alpha_{ij} = a_i$, $i = 1, 2, \ldots, m$ and $\sum_{j=0}^{m} \alpha_{0j} = a_0 - t$. We just observe that if $A_0 \cup A_1 \cup \cdots \cup A_m = \text{dom}(\theta)$ and $B_j = \{r \in \text{dom}(\theta) \cap \overline{A} : \theta(r) \in \Lambda_j\}$ then

$$[(\otimes_{w \in A_0} \phi_0) \otimes (\otimes_{w \in A_1} \phi_1) \otimes \cdots \otimes (\otimes_{w \in A_m} \phi_m)](\theta) = \prod_{i=0}^{m} \prod_{j=0}^{m} [\phi_i(j)]^{\alpha_{ij}},$$

where

$$\alpha_{ij} = |A_i \cap B_j| \qquad (3.31)$$

and for a fixed intersection matrix (α_{ij}) we have $\prod_{j=0}^{m} \binom{b_j}{\alpha_{0j}, \alpha_{1j}, \ldots, \alpha_{mj}}$ ways to choose the subsets $A_i \cap B_j$ of the B_j, and

$$A_0 = [\text{dom}(\theta) \setminus \overline{A}] \cap [\cup_{j=0}^{m}(A_0 \cap B_j)].$$

It follows that the value of $\Psi(n, h, \mathbf{a}, k)$ on a map θ with type$(\theta) = (t, \mathbf{b})$ is given by

$$\Psi(n, h, \mathbf{a}, k; t, \mathbf{b})$$

$$= \frac{1}{\binom{h}{a_0, a_1, \ldots, a_m}} \psi(n - \ell(\mathbf{a}), h - \ell(\mathbf{a}), k; h - t) \Phi(h, \mathbf{a}, h - t; \mathbf{b}).$$

Example 3.2.22 We specify Theorem 3.2.19 in the case where (F, H) is the Gelfand pair of the ultrametric space, that is, $Y = \{0, 1, \ldots, q - 1\}^m$, $F = \mathrm{Aut}(T_{q,m})$ and $H = K(q, m) \leq \mathrm{Aut}(T_{q,m})$ is the stabilizer of the leaf $y_0 = (0, 0, \ldots, 0)$. To simplify the notation, we assume that $2h \leq n$. Then Θ_h coincides with the space of all h-subsets $\{z_1, z_2, \ldots, z_h\}$ of the ultrametric space (Y, d) such that $d(z_i, z_j) = m$ (the maximum distance) for $i \neq j$.

From (2.15) it follows that, in this case, in (3.30) we have

$$\prod_{i=0}^{m} [\phi_i(j)]^{\alpha_{ij}}$$

$$= \begin{cases} \left(-\frac{1}{q-1} \right)^{\alpha_{m,1} + \alpha_{m-1,2} + \cdots + \alpha_{1,m}} & \text{if } \alpha_{i,j} = 0 \text{ for } i + j > m + 1 \\ 0 & \text{otherwise;} \end{cases}$$

that is, in (3.31) we must have $A_i \subseteq B_0 \cup B_1 \cup \cdots \cup B_{m-i+1}, i = 1, 2, \ldots, m$, and the value of $\prod_{i=0}^{m} [\phi_i(j)]^{\alpha_{ij}}$ is determined by the cardinalities $\gamma_j = |A_{m-j+1} \cap B_j|, j = 1, 2, \ldots, m$. Therefore

$$\Phi(h, \mathbf{a}, u; \mathbf{b})$$

$$= \sum_{\gamma} \prod_{j=1}^{m} \binom{b_j}{\gamma_j} \binom{\sum_{w=0}^{j-1} b_w - \sum_{v=1}^{j-1} a_{m-v+1}}{a_{m-j+1} - \gamma_j} \left(-\frac{1}{q-1} \right)^{\gamma_1 + \cdots + \gamma_m},$$

where the sum runs over all the m-tuples $\gamma = (\gamma_1, \gamma_2, \ldots, \gamma_m)$ satisfying

$$\max \left\{ 0, \sum_{v=1}^{j} a_{m-v+1} - \sum_{w=0}^{j-1} b_w \right\} \leq \gamma_j \leq \min\{b_j, a_{m-j+1}\}$$

(in particular, we have $\Phi(h, \mathbf{a}, t; \mathbf{b}) = 0$ when the conditions $\sum_{v=1}^{j} a_{m-v+1} \leq \sum_{w=0}^{j} b_w, j = 1, 2, \ldots, m - 1$, are not satisfied). So, to compute $\Phi(h, \mathbf{a}, u; k)$ we need to choose, in all possible ways,

- the subset $A_{m-j+1} \cap B_j$ in B_j, for $j = 1, 2, \ldots, m$,
- the subset $A_{m-j+1} \setminus B_j$ in $\left(\bigcup_{w=0}^{j-1} B_w \setminus \bigcup_{v=1}^{j-1} A_{m-v+1} \right)$, for $j = 1, 2, \ldots, m$,

and then, necessarily, $A_0 = \left[\bigcup_{w=0}^{m} B_w \setminus \bigcup_{v=1}^{m} A_{m-v+1} \right] \cup \left[\operatorname{dom}(\theta) \setminus \overline{A} \right]$.

Exercise 3.2.23 Deduce the results in [11, Section 7.4] as particular cases of those obtained in the present subsection.

3.2.6 The homogeneous space $\mathcal{V}(r, s)$ and the associated Gelfand pair

Let m be a positive integer and $\mathbf{r} = (r_1, r_2, \dots, r_m)$ an m-tuple of integers such that $r_i \geq 2$ for all $i = 1, 2, \dots, m$. In Section 2.1.4 we defined the spherically homogeneous rooted tree (**r**-tree) $T_{\mathbf{r}}$.

Let $\mathbf{s} = (s_1, s_2, \dots, s_m)$ be another m-tuple that satisfies the conditions $1 \leq s_i \leq r_i$, for all i. Denote by $T_{\mathbf{s}}$ the corresponding **s**-tree. Note that there are exactly

$$\binom{r_1}{s_1} \prod_{i=2}^{m} \binom{r_i}{s_i}^{s_1 s_2 \cdots s_{i-1}}$$

distinct embeddings of $T_{\mathbf{s}}$ as a subtree of $T_{\mathbf{r}}$. Indeed, any such subtree is uniquely determined by the m-tuple $(f_0, f_1, \dots, f_{m-1})$, where f_i is the map that associates with each vertex v at the ith level in $T_{\mathbf{s}}$ the set of all successors of v; for each i there are exactly $\binom{r_{i+1}}{s_{i+1}}^{s_1 s_2 \cdots s_i}$ such maps f_i.

We denote by $\mathcal{V}(\mathbf{r}, \mathbf{s})$ the set of all **s**-subtrees of $T_{\mathbf{r}}$ (see Fig. 3.3).

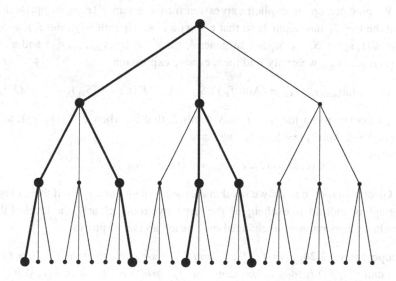

Fig. 3.3 A tree of type $(3, 3, 3)$ with a subtree of type $(2, 2, 1)$.

As usual, we denote by S_k the symmetric group on k elements and by $\mathrm{Aut}(T_\mathbf{r})$ the group of all automorphisms of $T_\mathbf{r}$; in Theorem 2.1.15 we showed that $\mathrm{Aut}(T_\mathbf{r}) = S_{r_m} \wr S_{r_{m-1}} \wr \cdots \wr S_{r_2} \wr S_{r_1}$. Recall from Section 2.1.4 that every automorphism $g \in \mathrm{Aut}(T_\mathbf{r})$ stabilizes the levels V_i of $T_\mathbf{r}$ and is uniquely determined by its labeling (cf. (2.11)), which we denote by $\bar{g}$. Thus, the map $\bar{g}$

$$V \to \bigcup_{i=0}^{m-1} S_{r_{i+1}}$$
$$v \mapsto \bar{g}(v)$$

(with $\bar{g}(v) \in S_{r_{i+1}}$ if $v \in V_i$) satisfies

$$g(x_1, x_2, \ldots, x_k) = (\bar{g}(\emptyset)x_1, \bar{g}(x_1)x_2, \ldots, \bar{g}(x_1, x_2, \ldots, x_{k-1})x_k)$$

for all $x_i \in X_i$, $1 \le i \le k \le m$. It is obvious that the image of an s-subtree under an automorphism of $T_\mathbf{r}$ is again a s-subtree. Thus, the group $\mathrm{Aut}(T_\mathbf{r})$ acts on $\mathcal{V}(\mathbf{r}, \mathbf{s})$.

Let now fix an s-subtree $T_\mathbf{s}^*$ and denote by

$$K(\mathbf{r}, \mathbf{s}) = \{g \in \mathrm{Aut}(T_\mathbf{r}) : gT_\mathbf{s}^* = T_\mathbf{s}^*\}$$

its stabilizer. We then have the identification

$$\mathcal{V}(\mathbf{r}, \mathbf{s}) = \mathrm{Aut}(T_\mathbf{r})/K(\mathbf{r}, \mathbf{s})$$

as $\mathrm{Aut}(T_\mathbf{r})$-spaces.

We now present an explicit expression for the group $K(\mathbf{r}, \mathbf{s})$. Suppose first that the tree $T_\mathbf{r}$ has depth 1, so that $\mathbf{r} = r_1$ and $\mathbf{s} = s_1$; clearly $\mathrm{Aut}(T_\mathbf{r}) = S_{r_1}$ and $K(\mathbf{r}, \mathbf{s}) = S_{s_1} \times S_{r_1-s_1}$. In general, for $\mathbf{r}' = (r_2, r_3, \ldots, r_m)$ and $\mathbf{s}' = (s_2, s_3, \ldots, s_m)$, we easily find the recursive expression

$$\mathrm{Stab}_{\mathrm{Aut}(T_\mathbf{r})}(T_\mathbf{s}) = (\mathrm{Aut}(T_{\mathbf{r}'}) \wr S_{r_1-s_1}) \times (K(\mathbf{r}', \mathbf{s}') \wr S_{s_1}). \tag{3.32}$$

In particular, when the tree $T_\mathbf{r}$ has depth 2, that is, when $\mathbf{r} = (r_1, r_2)$, $\mathbf{s} = (s_1, s_2)$ and $\mathrm{Aut}(T_\mathbf{r}) = S_{r_2} \wr S_{r_1}$, we have

$$K(\mathbf{r}, \mathbf{s}) = (S_{r_2} \wr S_{r_1-s_1}) \times ((S_{s_2} \times S_{r_2-s_2}) \wr S_{s_1}).$$

Given two rooted trees, we say that they are *rooted-isomorphic* if there exists a graph isomorphism exchanging the respective roots; clearly, the level of the single vertices remains unchanged under such an isomorphism.

Proposition 3.2.24 *Let T_1, T_2, T_1' and T_2' be s-subtrees within $T_\mathbf{r}$. Then (T_1, T_2) and (T_1', T_2') belong to the same $\mathrm{Aut}(T_\mathbf{r})$-orbit on $\mathcal{V}(\mathbf{r}, \mathbf{s}) \times \mathcal{V}(\mathbf{r}, \mathbf{s})$ if and only if $T_1 \cap T_2$ is rooted-isomorphic to $T_1' \cap T_2'$.*

Proof If $gT_j = T'_j$, $j = 1, 2$, for some $g \in \mathrm{Aut}(T_\mathbf{r})$ then $g(T_1 \cap T_2) = T'_1 \cap T'_2$ and the "only if" part follows trivially.

The other implication may be proved by induction on the depth m of the tree $T_\mathbf{r}$. For $m = 1$ we have $\mathbf{r} = r$ and $\mathbf{s} = s$, $\mathcal{V}(\mathbf{r}, \mathbf{s})$ is simply the set of all r-subsets of an s-set and $\mathrm{Aut}(T_\mathbf{r}) = S_r$; therefore we refer to the proof of Proposition 3.2.1.

Suppose now that $T_1 \cap T_2$ is rooted-isomorphic to $T'_1 \cap T'_2$ and denote by $\alpha : V_1(T_1 \cap T_2) \to V_1(T'_1 \cap T'_2)$ a bijection such that if $x \in V_1(T_1 \cap T_2)$ then the $(T_1 \cap T_2)$-subtree T_x rooted at x is (rooted-)isomorphic to the $(T'_1 \cap T'_2)$-subtree $T'_{\alpha(x)}$ rooted at $\alpha(x)$. We may extend α to a permutation $\sigma \in S_{r_1}$ such that $\sigma(V_1(T_1)) = V_1(T'_1)$ and $\sigma(V_1(T_2)) = V_1(T'_2)$. Modulo the permutation σ, we may suppose that $T_1 \cap T_2$ and $T'_1 \cap T'_2$ coincide at the first level.

By induction, for all $x \in V_1(T_1 \cap T_2) \equiv V_1(T'_1 \cap T'_2)$ we have an x-rooted isomorphism g_x between the $(T_1 \cap T_2)$-subtree rooted at x and the corresponding $(T'_1 \cap T'_2)$-subtree with the same root x. It is then clear that the automorphism g with label $\bar{g}(\emptyset) = \sigma$, $\bar{g}(x, x_2, \ldots, x_n) = \bar{g}_x(x_2, \ldots, x_n)$ if $x \in T_1 \cap T_2$ and the identity otherwise is the desired rooted automorphism. □

Corollary 3.2.25 *The action of* $\mathrm{Aut}(T_\mathbf{r})$ *on* $\mathcal{V}(\mathbf{r}, \mathbf{s})$ *is transitive.*

Proof It suffices to apply Proposition 3.2.24 to $T_1 = T_2$ and $T'_1 = T'_2$. □

Corollary 3.2.26 $(\mathrm{Aut}(T_\mathbf{r}), K(\mathbf{r}, \mathbf{s}))$ *is a symmetric Gelfand pair.*

Proof This follows from Proposition 3.2.24 by taking T_1, T_2, T'_1, T'_2 with $T'_1 = T_2$ and $T'_2 = T_1$ in combination with Proposition 1.2.31. □

Our next task is to relate the Gelfand pair $(\mathrm{Aut}(T_\mathbf{r}), K(\mathbf{r}, \mathbf{s}))$ to the generalized Johnson scheme. The classical Johnson scheme $(S_n, S_h \times S_{n-h})$ corresponds to the Gelfand pair $(\mathrm{Aut}(T_\mathbf{r}, T_\mathbf{s}), K(\mathbf{r}, \mathbf{s}))$, where $m = 1$, $\mathbf{r} = n$ and $\mathbf{s} = h$. More generally, given the Gelfand pair (F, H) with $F = \mathrm{Aut}(T_{\mathbf{r}'})$ and $H = K(\mathbf{r}', \mathbf{s}')$, $\mathbf{r}' = (r_2, r_3, \ldots, r_m)$ and $\mathbf{s}' = (s_2, s_3, \ldots, s_m)$, the homogeneous space Θ_h in Section 3.2.2 coincides with $\mathcal{V}(\mathbf{r}, \mathbf{s})$, where now $\mathbf{r} = (n, r_2, r_3, \ldots, r_m)$ and $\mathbf{s} = (h, s_2, s_3, \ldots, s_m)$. Indeed, the subgroup $(H \wr S_h) \times (F \wr S_{n-h})$ coincides with $K(\mathbf{r}, \mathbf{s})$ by virtue of the expression given in (3.32). The point stabilized by $(H \wr S_h) \times (F \wr S_{n-h})$, namely $\theta_0 \in \Theta_h$ (which corresponds to an h-subset $\overline{A} \subset \{1, 2, \ldots, n\}$), is given by $\theta_0(j) = y_0$ for all $j \subset \overline{A}$, where y_0 is the $\mathbf{s}'$-subtree stabilized by $H \equiv K(\mathbf{r}', \mathbf{s}')$.

Remark 3.2.27 In Example 3.2.22 we considered the Gelfand pair (F, H) (the ultrametric space), where $F = \mathrm{Aut}(T_{q,m})$ and $H = K(q, m) \leq \mathrm{Aut}(T_{q,m})$ is the stabilizer of the leaf $y_0 = (0, 0, \ldots, 0) \in Y = \{0, 1, \ldots, q-1\}^m$. In the setting of the present section, the corresponding homogeneous space Θ_h

coincides with the homogeneous space $\mathcal{V}(\mathbf{r},\mathbf{s})$, with $\mathbf{r} = (n, q, \ldots, q)$ and $\mathbf{s} = (h, 1, \ldots, 1)$.

3.3 Harmonic analysis on exponentiations and on wreath products of permutation representations

In this section we introduce the notion of a wreath product of permutation representations that generalizes the exponentiation action (cf. (2.7)), the Cayley action of a wreath product on itself and the lamplighter space (which will be examined in more detail in Section 3.4). We give explicit rules for the decomposition of the corresponding permutation representations into irreducibles and analyze several special cases.

3.3.1 Exponentiation and wreath products

Definition 3.3.1 Let G (resp. F) be a finite group acting on two finite sets X and Z (resp. on a finite set Y). Then we define an action of the wreath product $F \wr G = F^X \rtimes G$ on $Y^X \times Z$ by setting

$$(f, g)(\varphi, z) = ((f, g)\varphi, gz) \tag{3.33}$$

for all $(f, g) \in F \wr G$ and $(\varphi, z) \in Y^X \times Z$. We will call it the *wreath product* of the action of F on Y and the actions of G on X and Z.

Note that (3.33) is just the direct product of the exponentiation action (of $F \wr G$ on Y^X, see (2.7)) with the *inflation* of the action of G on Z (which is defined by setting $(f, g)z = gz$ for $(f, g) \in F \wr G$ and $z \in Z$). In particular, if the action of G on Z and the action of F on Y are transitive then the wreath product (3.33) is also a transitive action: recall that, in this case, F^X is transitive on Y^X (see Lemma 2.1.8). When Z is trivial, (3.33) coincides with the exponentiation action.

Exercise 3.3.2 Suppose that $Y = F$ and $Z = G$. Show that the wreath product of the left Cayley action of F on itself, the action of G on X and the left Cayley action of G on itself coincides with the left Cayley action of $F \wr G$ on itself.

Notation 3.3.3 We will use the following notation for the permutation representation of F on Y: if $\xi \in L(Y)$ and $f \in F$, we set $(f\xi)(y) = \xi(f^{-1}y)$ for all $y \in Y$. We will use a similar notation for the permutation representation of G on Z. However, we will denote by λ the permutation representation of F^X

on $L\left(Y^X\right)$ and by $\widetilde{\lambda}$ its extension to $F \wr G$ (which coincides with the permutation representation associated with exponentiation). That is, for $\psi \in L\left(Y^X\right)$ and $f \in F^X$ we have

$$[\lambda(f)\psi](\varphi) = \psi(f^{-1}\varphi) \qquad \text{for all } \varphi \in Y^X,$$

where $(f^{-1}\varphi)(x) = f(x)^{-1}\varphi(x)$ for all $x \in X$. Similarly, if in addition $g \in G$, we have

$$[\widetilde{\lambda}(f, g)\psi](\varphi) = \psi[(f, g)^{-1}\varphi] \equiv \psi[(g^{-1}f^{-1}, g^{-1})\varphi],$$

where

$$[(g^{-1}f^{-1}, g^{-1})\varphi](x) = f(gx)^{-1}\varphi(gx)$$

for all $\varphi \in Y^X$ and $x \in X$. In particular $\left[\widetilde{\lambda}(1_F, g)\psi\right](\varphi) = \psi(g^{-1}\varphi)$, where $(g^{-1}\varphi)(x) = \varphi(gx)$. Finally, we will use the following notation for the permutation representation of $F \wr G$ on $L\left(Y^X \times Z\right)$: if $\Psi \in L\left(Y^X \times Z\right)$ and $(f, g) \in F \wr G$, we set

$$[(f, g)\Psi](\varphi, z) = \Psi[(f, g)^{-1}(\varphi, z)] \qquad \text{for all } (\varphi, z) \in Y^X \times Z.$$

Now let $\psi_x \in L(Y)$ for all $x \in X$. We define the tensor product $\bigotimes_{x \in X} \psi_x \in L\left(Y^X\right)$ by setting

$$\left(\bigotimes_{x \in X} \psi_x\right)(\varphi) = \prod_{x \in X} \psi_x(\varphi(x)) \qquad \text{for all } \varphi \in Y^X. \tag{3.34}$$

Compare with (3.2); in particular, λ may be seen as the $|X|$-times tensor product of the permutation representation of F on Y. Therefore, the following lemma may be considered as a particular case of Lemma 2.4.3, just noting that the inertia group of λ coincides with G. However, we give an easy proof for the reader's convenience.

Lemma 3.3.4 *Let $\psi_x \in L(Y)$, $x \in X$, and $(f, g) \in F \wr G$. Then*

$$\widetilde{\lambda}(f, g)\left(\bigotimes_{x \in X} \psi_x\right) = \bigotimes_{x \in X} f(x)\psi_{g^{-1}x}.$$

Proof For any $\varphi \in Y^X$ and $x \in X$, by (2.7) we have $[(f, g)^{-1}\varphi](x) = f(gx)^{-1}\varphi(x)$ and therefore

$$\left[\tilde{\lambda}(f,g)\left(\bigotimes_{x\in X}\psi_x\right)\right](\varphi) = \left(\bigotimes_{x\in X}\psi_x\right)[(f,g)^{-1}\varphi]$$

$$= \prod_{x\in X}\psi_x[f(gx)^{-1}\varphi(gx)] \qquad \text{(by (3.34))}$$

$$= \prod_{x\in X}\psi_{g^{-1}x}[f(x)^{-1}\varphi(x)]$$

$$= \prod_{x\in X}\left[f(x)\psi_{g^{-1}x}\right](\varphi(x))$$

$$= \left[\bigotimes_{x\in X}f(x)\psi_{g^{-1}x}\right](\varphi). \qquad \square$$

Let Σ be a system of representatives for the $(F \wr G)$-conjugacy classes of $\widehat{F^X}$ (as in Theorem 2.4.4), fix $\sigma \in \Sigma$ and set $I = T_G(\sigma)$. In the notation of Section 2.4, we also assume that for all $x \in X$ the representation σ_x appears in the decomposition of $L(Y)$ into irreducible F-representations. Then there exists a partition $X = \coprod_{i=1}^{n}\Omega_i$ of X and $\sigma_1, \sigma_2, \ldots, \sigma_n$ irreducible and pairwise inequivalent F-representations such that $\sigma_x = \sigma_i$ for all $x \in \Omega_i, i = 1, \ldots, n$, and $I = \{g \in G : g\Omega_i = \Omega_i, i = 1, 2, \ldots, n\}$. Denote by m_i the multiplicity of σ_i in $L(Y)$. For each $x \in \Omega_i$ we denote by $V_{\sigma_x} \equiv V_i$ the representation space of $\sigma_x \equiv \sigma_i$. We fix a basis $T_{x,1}, T_{x,2}, \ldots, T_{x,m_i}$ in $\text{Hom}_F(V_{\sigma_x}, L(Y))$ which is orthonormal with respect to the Hilbert–Schmidt scalar product. This means that (cf. (1.35))

$$\text{Tr}\left[T_{x,h}\left(T_{x,k}\right)^*\right] \equiv d_{\sigma_i}\langle T_{x,h}, T_{x,k}\rangle_{\text{HS}} = \delta_{h,k}$$

for all $1 \leq h, k \leq m_i$. Moreover we suppose that, for all $x \in \Omega_i$, the operators $T_{x,h} : V_i \to L(Y), 1 \leq h \leq m_i$, are all the same. We set

$$J = \{j \in \mathbb{N}^X : 1 \leq j(x) \leq m_i \quad \text{for all } x \in \Omega_i, i = 1, 2, \ldots, n\}. \qquad (3.35)$$

The group I acts on J in the obvious way: if $g \in I$ and $j \in J$ then gj is defined by setting $gj(x) = j(g^{-1}x)$ for all $x \in X$ (recall that I stabilizes every Ω_i). For any $j \in J$, set

$$T_j = \bigotimes_{x\in X}T_{x,j(x)}. \qquad (3.36)$$

It is easy to check that $\{T_j : j \in J\}$ is an orthonormal basis for

$$\text{Hom}_{F^X}\left(\bigotimes_{x\in X}V_{\sigma_x}, L\left(Y^X\right)\right).$$

We also have

$$T_{gx,j(gx)} = T_{x,j(gx)} \qquad \text{if } g \in I, \qquad (3.37)$$

because $T_{x,h}$ does not depend on $x \in \Omega_i$. Recall that, in the notation of Lemma 2.4.3, $\widetilde{\sigma}$ is the extension of σ to $F \wr I$.

Lemma 3.3.5   *For $g \in I$ and $T \in \mathrm{Hom}_{FX}\left(\bigotimes_{x \in X} V_{\sigma_x}, L\left(Y^X\right)\right)$, define a linear operator $\pi(g)T : V_{\sigma_x} \to L\left(Y^X\right)$ by setting*

$$\pi(g)T = \widetilde{\lambda}(\mathbf{1}_F, g) T \widetilde{\sigma}(\mathbf{1}_F, g^{-1}).$$

Then π is a representation of I on $\mathrm{Hom}_{FX}\left(\bigotimes_{x \in X} V_{\sigma_x}, L\left(Y^X\right)\right)$.

Proof   We first check that

$$\widetilde{\lambda}(\mathbf{1}_F, g) T \widetilde{\sigma}(\mathbf{1}_F, g^{-1}) \in \mathrm{Hom}_{FX}\left(\bigotimes_{x \in X} V_{\sigma_x}, L\left(Y^X\right)\right)$$

for every $T \in \mathrm{Hom}_{FX}\left(\bigotimes_{x \in X} V_{\sigma_x}, L\left(Y^X\right)\right)$ and $g \in I$. Noting that $\sigma(f, 1_G) = \widetilde{\sigma}(f, 1_G)$ and

$$(\mathbf{1}_F, g^{-1})(f, 1_G) = (g^{-1}f, g^{-1}) = (g^{-1}f, 1_G)(\mathbf{1}_F, g^{-1}),$$

we have

$$\left[\widetilde{\lambda}(\mathbf{1}_F, g) T \widetilde{\sigma}(\mathbf{1}_F, g^{-1})\right]\sigma(f, 1_G) = \widetilde{\lambda}(\mathbf{1}_F, g) T \widetilde{\sigma}(g^{-1}f, 1_G)\widetilde{\sigma}(\mathbf{1}_F, g^{-1})$$

$$= \widetilde{\lambda}(\mathbf{1}_F, g)\widetilde{\lambda}(g^{-1}f, 1_G) T \widetilde{\sigma}(\mathbf{1}_F, g^{-1})$$

$$= \widetilde{\lambda}(f, 1_G)\left[\widetilde{\lambda}(\mathbf{1}_F, g) T \widetilde{\sigma}(\mathbf{1}_F, g^{-1})\right].$$

We leave to the reader the easy verification that π is a representation.   □

Lemma 3.3.6   *For all $(f, g) \in F \wr I$ and $j \in J$ we have*

$$\widetilde{\lambda}(f, g)T_j = T_{gj}\widetilde{\sigma}(f, g).$$

Proof   We examine the actions of $\widetilde{\lambda}(f, g)T_j$ and $T_{gj}\widetilde{\sigma}(f, g)$ on the tensor product $\bigotimes_{x \in X} v_x \in \bigotimes_{x \in X} V_{\sigma_x}$ separately, showing that they lead to the same expression. From (3.36), Lemma 3.3.4 and (3.37), we deduce that

$$\left[\widetilde{\lambda}(f, g)T_j\right]\left(\bigotimes_{x \in X} v_x\right) = \widetilde{\lambda}(f, g)\left(\bigotimes_{x \in X} T_{x,j(x)}v_x\right)$$

$$= \bigotimes_{x \in X} f(x)T_{x,j(g^{-1}x)} v_{g^{-1}x}.$$

Similarly, from the definition of $\widetilde{\sigma}$ in Lemma 2.4.3, taking into account that $\sigma_{g^{-1}x} = \sigma_x$ for $g \in I$ and the fact that $T_{x,j(x)} \in \mathrm{Hom}_F(V_{\sigma_x}, L(Y))$, $x \in X$, it follows that

$$
\left[T_{gj}\widetilde{\sigma}(f,g)\right]\left(\bigotimes_{x \in X} v_x\right) = \left(\bigotimes_{x \in X} T_{x,gj(x)}\right)\left(\bigotimes_{x \in X} \sigma_x(f(x))v_{g^{-1}x}\right)
$$
$$
= \bigotimes_{x \in X} T_{x,gj(x)}\sigma_x(f(x))v_{g^{-1}x}
$$
$$
= \bigotimes_{x \in X} f(x)T_{x,j(g^{-1}x)}v_{g^{-1}x}.
$$

$\square$

Corollary 3.3.7 *The representation π in Lemma 3.3.5 is equivalent to the permutation representation of the group I on the finite set J.*

Proof Let $g \in I$ and $j \in J$. On the one hand, setting $f = \mathbf{1}_F$ in Lemma 3.3.6 we get the identity

$$
\pi(g)T_j \equiv \widetilde{\lambda}(\mathbf{1}_F, g)T_j\widetilde{\sigma}(\mathbf{1}_F, g^{-1}) = T_{gj}.
$$

On the other hand, we also have $g\delta_j = \delta_{gj}$. Since $\{T_j : j \in J\}$ (resp. $\{\delta_j : j \in J\}$) is an orthonormal basis for $\mathrm{Hom}_{FX}\left(\bigotimes_{x \in X} V_{\sigma_x}, L\left(Y^X\right)\right)$ (resp. for $L(J)$), we conclude that the map

$$
L(J) \longrightarrow \mathrm{Hom}_{FX}\left(\bigotimes_{x \in X} V_{\sigma_x}, L\left(Y^X\right)\right)
$$
$$
\delta_j \longmapsto T_j
$$

is an isomorphism of I-representations. $\square$

Now let (η, U) be an irreducible representation of I. Recall that the inflation $\overline{\eta}$ of η to $F \wr I$ is defined by

$$
\overline{\eta}(f,g)u = \eta(g)u \qquad \text{for all } (f,g) \in F \wr I \text{ and } u \in U. \tag{3.38}
$$

The key point is to determine the multiplicity of the representation

$$
\left(\widetilde{\sigma} \otimes \overline{\eta}, \left(\bigotimes_{x \in X} V_{\sigma_x}\right) \otimes U\right)
$$

in the decomposition of $L\left(Y^X \times Z\right)$ into irreducible $(F \wr I)$-representations. In the spirit of the harmonic analysis developed in Section 1.2, what we actually get is an orthogonal decomposition which depends on an orthogonal decomposition of the permutation representation of I on $L(J \times Z)$. For $g \in I$, we denote by $g\gamma$ the g-image of $\gamma \in L(J \times Z)$.

Theorem 3.3.8 *Let $\sigma \in \Sigma$ and let (η, U) be an irreducible representation of $I = T_G(\sigma)$. Let $\{T_j : j \in J\}$ be an orthonormal basis for $\mathrm{Hom}_{FX}\left(\bigotimes_{x \in X} V_{\sigma_x}, L(Y^X)\right)$ and $T \in \mathrm{Hom}_I(U, L(J \times Z))$. Given*

$$v = \bigotimes_{x \in X} v_x \in \bigotimes_{x \in X} V_{\sigma_x}$$

and $u \in U$, define $\widehat{T}(v \otimes u) \in L\left(Y^X \times Z\right)$ by setting

$$\left[\widehat{T}(v \otimes u)\right](\varphi, z) = \left[\sum_{j \in J}(Tu)(j, z)T_j v\right](\varphi) \equiv \sum_{j \in J}(Tu)(j, z) \cdot (T_j v)(\varphi)$$

$$(3.39)$$

for all $(\varphi, z) \in Y^X \times Z$. Then the following hold:

(i) *The operator $\widehat{T}$ belongs to*

$$\mathrm{Hom}_{F \wr I}\left[\left(\bigotimes_{x \in X} V_{\sigma_x}\right) \otimes U, L\left(Y^X \times Z\right)\right].$$

(ii) *The map*

$$\mathrm{Hom}_I(U, L(Z \times J)) \quad \rightarrow \quad \mathrm{Hom}_{F \wr I}\left[\left(\bigotimes_{x \in X} V_{\sigma_x}\right) \otimes U, L\left(Y^X \times Z\right)\right]$$

$$T \qquad\qquad \longmapsto \qquad\qquad \widehat{T}$$

is a linear isometric isomorphism.

Proof   Recall that $\widetilde{\sigma}$ (resp. $\overline{\eta}$) is the extension (resp. inflation) of σ (resp. η) to $F \wr I$. Let $(f, g) \in F \wr I$ and $(\varphi, z) \in Y^X \times Z$. Then

$$\left(\widehat{T}\{[\widetilde{\sigma}(f, g) \otimes \overline{\eta}(f, g)](v \otimes u)\}\right)(\varphi, z)$$

$$= \left(\widehat{T}\left[\widetilde{\sigma}(f, g)v \otimes \eta(g)u\right]\right)(\varphi, z) \qquad \text{(by (3.38))}$$

$$- \left(\sum_{j \in J}\{[T\eta(g)u](j, z)\} \cdot T_j\widetilde{\sigma}(f, g)v\right)(\varphi) \qquad \text{(by (3.39))}$$

$$= \left(\widetilde{\lambda}(f, g)\left(\sum_{j \in J}[T(u)(g^{-1}j, g^{-1}z)] \cdot T_{g^{-1}j}v\right)\right)(\varphi)$$

(since $\mathcal{T} \in \mathrm{Hom}_I(U, L(J \times Z))$ and $\tilde{\lambda}(f,g)T_{g^{-1}j} = T_j\tilde{\sigma}(f,g)$ by Lemma 3.3.6)

$$= \left(\sum_{j\in J} [\mathcal{T}(u)(j, g^{-1}z)] \cdot T_j v \right) ((f,g)^{-1}\varphi) \qquad \text{(on replacing } g^{-1}j \text{ by } j\text{)}$$

$$= [\widehat{\mathcal{T}}(v \otimes u)]((f,g)^{-1}\varphi, g^{-1}z) \qquad \text{(again by (3.39))}$$

$$= ((f,g)[\widehat{\mathcal{T}}(v \otimes u)])(\varphi, z).$$

This proves that $\widehat{\mathcal{T}}$ is an intertwiner.

In order to show that the correspondence $\mathcal{T} \longmapsto \widehat{\mathcal{T}}$ is a bijection, we construct an explicit inverse map $\widehat{\mathcal{T}} \longmapsto \mathcal{T}$ as follows. Let

$$S \in \mathrm{Hom}_{F?I}\left(\left(\bigotimes_{x\in X} V_{\sigma_x} \right) \otimes U, L\left(Y^X \times Z\right) \right). \qquad (3.40)$$

With every choice of $u \in U$ and $z \in Z$ we associate a linear map

$$S_{u,z}^\sharp : \bigotimes_{x\in X} V_{\sigma_x} \longrightarrow L\left(Y^X\right),$$

defined by setting

$$(S_{u,z}^\sharp v)(\varphi) = [S(v \otimes u)](\varphi, z) \qquad (3.41)$$

for all $v = \bigotimes_{x\in X} v_x \in \bigotimes_{x\in X} V_{\sigma_x}$ and $\varphi \in Y^X$. Let us show that

$$S_{u,z}^\sharp \in \mathrm{Hom}_{FX}\left(\bigotimes_{x\in X} V_{\sigma_x}, L\left(Y^X\right) \right). \qquad (3.42)$$

For all $f \in F^X$ and $\varphi \in Y^X$ we have:

$$\left[S_{u,z}^\sharp \sigma(f)v \right](\varphi)$$

$$= \left[S_{u,z}^\sharp \tilde{\sigma}(f, 1_G)v \right](\varphi)$$

$$= \left\{ S\left[(\tilde{\sigma}(f, 1_G)v) \otimes u \right] \right\}(\varphi, z)$$

$$= (S[\tilde{\sigma}(f, 1_G) \otimes \overline{\eta}(f, 1_G)](v \otimes u))(\varphi, z) \qquad \text{(since } \overline{\eta}(f, 1_G)u = u\text{)}$$

$$= [S(v \otimes u)]\left((f, 1_G)^{-1}\varphi, z\right) \qquad \text{(by (3.40))}$$

$$= (S_{u,z}^\sharp v)(f^{-1}\varphi)$$

$$= [\lambda(f)S_{u,z}^\sharp v](\varphi),$$

that is,

$$S_{u,z}^\sharp \sigma(f) = \lambda(f)S_{u,z}^\sharp.$$

Thus (3.42) is proved. Since the operators $\{T_j \, : \, j \in J\}$ form a basis for $\text{Hom}_{FX}\left(\bigotimes_{x \in X} V_{\sigma_x}, L\left(Y^X\right)\right)$, we deduce that there exist $\alpha_{u,z} \in L(J)$ such that

$$S^{\sharp}_{u,z} = \sum_{j \in J} \alpha_{u,z}(j)T_j. \tag{3.43}$$

Therefore we define a linear map $\widetilde{S} : U \rightarrow L(J \times Z)$ by setting, for $u \in U$ and $(j, z) \in J \times Z$,

$$\left(\widetilde{S}u\right)(j, z) = \alpha_{u,z}(j), \tag{3.44}$$

where $\alpha_{u,z}$ is given by (3.43). From (3.41), (3.43) and (3.44) it follows that

$$[S\,(v \otimes u)]\,(\varphi, z) = \left[\sum_{j \in J} \alpha_{u,z}(j)\,T_j v\right](\varphi) = \left[\sum_{j \in J} \left(\widetilde{S}u\right)(j, z)\,T_j v\right](\varphi). \tag{3.45}$$

Moreover, for all $g \in I$ we have

$$\left(\sum_{j \in J} \left(\left[\widetilde{S}\eta(g)u\right](j, z)\right)\,T_j v\right)(\varphi)$$

$$= \left(S\left[\widetilde{\sigma}(1_F, g)\widetilde{\sigma}(1_F, g^{-1})v \otimes \eta(g)u\right]\right)(\varphi, z)$$

$$= \left(S\left[\widetilde{\sigma}(1_F, g^{-1})v \otimes u\right]\right)\left((1_F, g^{-1})\varphi, (1_F, g)^{-1}z\right) \qquad \text{(by (3.40))}$$

$$= \left[\sum_{j \in J} \left(\widetilde{S}u\right)(j, g^{-1}z)\,T_j\widetilde{\sigma}(1_F, g^{-1})v\right]\left[(1_F, g^{-1})\varphi\right]$$

$$= \left[\sum_{j \in J} \left(\widetilde{S}u\right)(j, g^{-1}z)\,T_{gj}v\right](\varphi) \qquad \text{(by Lemma 3.3.6)}$$

$$= \left[\sum_{j \in J} \left(\widetilde{S}u\right)(g^{-1}j, g^{-1}z)\,T_j v\right](\varphi)$$

$$= \left[\sum_{j \in J} \left(g\widetilde{S}u\right)(j, z)\,T_j v\right](\varphi).$$

This shows that $\widetilde{S}\eta(g) = g\widetilde{S}$, that is, $\widetilde{S} \in \text{Hom}_I(U, L(J \times Z))$. From (3.39) and (3.45), it is also clear that $\widetilde{\widetilde{S}} = S$ and that $\widetilde{\widetilde{T}} = T$, thus showing the bijectivity of the map $T \longmapsto \widehat{T}$.

Finally, we end the proof by showing that the above map is indeed an isometry. Let $\mathcal{B}_1$ (resp. $\mathcal{B}_2$) be an orthonormal basis in $\bigotimes_{x \in X} V_{\sigma_x}$ (resp. U). If $\mathcal{T}_1, \mathcal{T}_2 \in \mathrm{Hom}_I(U, L(J \times Z))$, then

$$
\langle \widehat{\mathcal{T}_1}, \widehat{\mathcal{T}_2} \rangle_{\mathrm{HS}} = \sum_{v \in \mathcal{B}_1} \sum_{u \in \mathcal{B}_2} \sum_{(\varphi, z) \in Y^X \times Z} [\widehat{\mathcal{T}_1}(v \otimes u)](\varphi, z) \overline{[\widehat{\mathcal{T}_2}(v \otimes u)](\varphi, z)}
$$

$$
= \sum_{v, u} \sum_{(\varphi, z)} \left[\sum_{j \in J} (\mathcal{T}_1 u)(j, z)(T_j v)(\varphi) \right] \left[\sum_{i \in J} \overline{(\mathcal{T}_2 u)(i, z)(T_i v)(\varphi)} \right]
$$

$$
= \sum_{u, j, i, z} (\mathcal{T}_1 u)(j, z) \overline{(\mathcal{T}_2 u)(i, z)} \langle T_j, T_i \rangle_{\mathrm{HS}}
$$

$$
= \sum_{u, j, z} (\mathcal{T}_1 u)(j, z) \overline{(\mathcal{T}_2 u)(j, z)}
$$

$$
= \langle \mathcal{T}_1, \mathcal{T}_2 \rangle_{\mathrm{HS}}. \hspace{6cm} \square
$$

Exercise 3.3.9 Show that the linear map $\mathcal{T}^\sharp : U \otimes L(Z) \to L(J)$ defined by $\mathcal{T}^\sharp(u \otimes \delta_z) = \mathcal{T}^\sharp_{u,z}$ belongs to $\mathrm{Hom}_I(U \otimes L(Z), L(J))$ and that

$$
\mathrm{Hom}_{F \wr I} \left(\left(\bigotimes_{x \in X} V_{\sigma_x} \right) \otimes U, L\left(Y^X \times Z \right) \right) \longrightarrow \mathrm{Hom}_I(U \otimes L(Z), L(J))
$$
$$
\mathcal{T} \longmapsto \mathcal{T}^\sharp
$$

is a linear isomorphism.
Hint. Recall Corollary 3.3.7.

The formulation of Frobenius reciprocity in Theorem 1.2.27 yields an explicit isometric isomorphism from

$$
\mathrm{Hom}_{F \wr I} \left(\left(\bigotimes_{x \in X} V_{\sigma_x} \right) \otimes U, L\left(Y^X \times Z \right) \right)
$$

onto

$$
\mathrm{Hom}_{F \wr G} \left(\mathrm{Ind}_{F \wr I}^{F \wr G} \left[\left(\bigotimes_{x \in X} V_{\sigma_x} \right) \otimes U \right], L\left(Y^X \times Z \right) \right),
$$

given by

$$
\mathcal{T} \longmapsto \overset{\diamond}{\mathcal{T}}. \hspace{5cm} (3.46)
$$

By combining Theorem 3.3.8 and the isomorphism (3.46), we can reduce the decomposition of $L\left(Y^X \times Z \right)$ into irreducible $(F \wr G)$-representations to the decomposition of $L(J \times Z)$ into irreducible I-representations.

Theorem 3.3.10 *The map* $\mathcal{T} \longmapsto \overset{\diamond}{\widehat{\mathcal{T}}}$ *from* $\mathrm{Hom}_I(U, L(Z \times J))$ *into*

$$\mathrm{Hom}_{F \wr G}\left(\mathrm{Ind}_{F \wr I}^{F \wr G}\left[\left(\bigotimes_{x \in X} V_{\sigma_x}\right) \otimes U\right], L\left(Y^X \times Z\right)\right)$$

is a linear isometric isomorphism. □

The following results are an immediate consequence of the above.

Corollary 3.3.11 *The multiplicity of the irreducible representation* $\mathrm{Ind}_{F \wr I}^{F \wr G}$ $(\widetilde{\sigma} \otimes \overline{\eta})$ *in the decomposition of* $L\left(Y^X \times Z\right)$ *into irreducible* $(F \wr G)$- *representations is equal to the multiplicity of* η *in the decomposition of* $L(J \times Z)$ *into irreducible* I-*representations.* □

In particular, if Z is trivial then we get a rule for the decomposition of the exponentiation action into irreducible representations:

Corollary 3.3.12 *The multiplicity of* $\mathrm{Ind}_{F \wr I}^{F \wr G}(\widetilde{\sigma} \otimes \overline{\eta})$ *in* $L\left(Y^X\right)$ *is equal to the multiplicity of* η *in* $L(J)$. □

Remark 3.3.13 If we take $Y \equiv F$ and $Z \equiv G$, both with their left Cayley actions, then Theorem 3.3.10 yields a decomposition of the left regular representation of $F \wr G$ (see Exercise 3.3.2). Since the stabilizer in I of any $(j, g) \in J \times G$ is the trivial subgroup, each orbit of I on $J \times G$ is equivalent to the left action of I on itself and therefore I has $|J| \frac{|G|}{|I|}$ orbits on $J \times G$. Hence the multiplicity η in $L(J \times G)$ is equal to $(\dim \eta) |J| \frac{|G|}{|I|}$; by virtue of Corollary 3.3.11, this yields a formula for $\dim \mathrm{Ind}_{F \wr I}^{F \wr G}(\widetilde{\sigma} \otimes \overline{\eta})$. Note that this agrees with (1.5), since then $\dim \mathrm{Ind}_{F \wr I}^{F \wr G}(\widetilde{\sigma} \otimes \overline{\eta}) = (\dim \eta)(\dim \sigma)\frac{|G|}{|I|}$ (in this context we have $\dim \sigma = |J|$).

The results in the following subsections are all particular cases and applications of Theorem 3.3.8 and Corollary 3.3.11. But it is worthwhile to examine them separately, explore their peculiarities and develop more direct approaches when possible.

3.3.2 The case $G = C_2$ and Z trivial

In the present subsection we examine the case $G = C_2 \equiv X$ (with the Cayley action) and Z trivial. We identify C_2 with the multiplicative group $\{1, -1\}$ and denote by (ε, U) the corresponding alternating representation. As before, we also consider a finite group F acting on a finite set Y, and we denote by

$$L(Y) = \bigotimes_{\sigma \in R} m_\sigma V_\sigma \qquad (3.47)$$

the isotypic decomposition of the corresponding permutation representation into irreducible F-representations. Finally, let $T_1^\sigma, T_2^\sigma, \ldots, T_{m_\sigma}^\sigma$ constitute an orthonormal basis for $\mathrm{Hom}_F(V_\sigma, L(Y))$. We clearly have that

$$L(Y \times Y) \equiv L(Y) \bigotimes L(Y) = \bigoplus_{\sigma, \sigma' \in R} m_\sigma m_{\sigma'} (V_\sigma \otimes V_{\sigma'})$$

$$\equiv \bigoplus_{\sigma, \sigma' \in R} \bigoplus_{j=1}^{m_\sigma} \bigoplus_{j'=1}^{m_{\sigma'}} \left(T_j^\sigma V_\sigma \otimes T_{j'}^{\sigma'} V_{\sigma'} \right) \qquad (3.48)$$

is an orthogonal decomposition of $L(Y \times Y)$ into irreducible $(F \times F)$-representations.

Theorem 3.3.14 *We will use the notation in (3.47).*

(i) *For $\sigma, \sigma' \in R$, $\sigma \neq \sigma'$, $j = 1, 2, \ldots, m_\sigma$ and $j' = 1, 2, \ldots, m_{\sigma'}$, the space*

$$W_{\sigma, \sigma'}^{j, j'} = \left(T_j^\sigma V_\sigma \otimes T_{j'}^{\sigma'} V_{\sigma'} \right) \oplus \left(T_{j'}^{\sigma'} V_{\sigma'} \otimes T_j^\sigma V_\sigma \right)$$

is an irreducible $(F \wr C_2)$-representation isomorphic to $(V_\sigma \otimes V_{\sigma'}) \oplus (V_{\sigma'} \otimes V_\sigma)$.

(ii) *For $\sigma \in R$ and $i, j = 1, 2, \ldots, m_\sigma$, the space*

$$W_{\sigma, +}^{i, j} = \left\langle \left(T_i^\sigma v_1 \otimes T_j^\sigma v_2 \right) + \left(T_j^\sigma v_1 \otimes T_i^\sigma v_2 \right) : v_1, v_2 \in V_\sigma \right\rangle$$

is an irreducible $(F \wr C_2)$-representation isomorphic to $V_\sigma \otimes V_\sigma$.

(iii) *For $\sigma \in R$ and $i, j = 1, 2, \ldots, m_\sigma$, $i \neq j$, the space*

$$W_{\sigma, -}^{i, j} = \left\langle \left(T_i^\sigma v_1 \otimes T_j^\sigma v_2 \right) - \left(T_j^\sigma v_1 \otimes T_i^\sigma v_2 \right) : v_1, v_2 \in V_\sigma \right\rangle$$

is an irreducible $(F \wr C_2)$-representation isomorphic to $(V_\sigma \otimes V_\sigma) \otimes U$.

(iv) *The decomposition of $L(Y \times Y)$ into irreducible $(F \wr C_2)$-representations is given by*

$$L(Y \times Y) \cong \left[\bigoplus_{\substack{\sigma, \sigma' \in R \\ \sigma \neq \sigma'}} m_\sigma m_{\sigma'} (V_\sigma \otimes V_{\sigma'}) \oplus (V_{\sigma'} \otimes V_\sigma) \right]$$

$$\oplus \left[\bigoplus_{\sigma \in R} \frac{m_\sigma (m_\sigma + 1)}{2} (V_\sigma \otimes V_\sigma) \right]$$

$$\oplus \left[\bigoplus_{\sigma \in R} \frac{m_\sigma (m_\sigma - 1)}{2} [(V_\sigma \otimes V_\sigma) \otimes U] \right]$$

$$\cong \left(\bigoplus_{\sigma, \sigma' \in R\sigma \neq \sigma'} \bigoplus_{i=1}^{m_\sigma} \bigoplus_{j=1}^{m_{\sigma'}} W_{\sigma, \sigma'}^{j, j'} \right)$$

$$\oplus \left(\bigoplus_{\sigma \in R} \left[\left(\bigoplus_{i, j = 1}^{m_\sigma} W_{\sigma, +}^{i, j} \right) \oplus \left(\bigoplus_{\substack{i, j = 1 \\ i \neq j}}^{m_\sigma} W_{\sigma, -}^{i, j} \right) \right] \right).$$

Proof When $\sigma \neq \sigma'$, the inertia group of $V_\sigma \otimes V_{\sigma'}$ is $F \times F$. Indeed, $T_G(\sigma \otimes \sigma')$ is trivial (since $\sigma \neq \sigma'$) and therefore (see Lemma 2.4.2) $I_{F^X \wr G}(\sigma \otimes \sigma') = F \wr T_G(\sigma \otimes \sigma') = F^X = F \times F$. Moreover, from Proposition 1.1.9 it follows that

$$\text{Ind}_{F \times F}^{F \wr C_2} (V_\sigma \otimes V_{\sigma'}) \cong (V_\sigma \otimes V_{\sigma'}) \oplus (V_{\sigma'} \otimes V_\sigma).$$

From Theorem 2.4.4 we deduce that this is an irreducible $(F \wr C_2)$-representation, and (3.48) ensures that its multiplicity in $L(Y \times Y)$ is equal to $m_\sigma m_{\sigma'}$. This agrees with Corollary 3.3.12 because, with the notation therein, we have that $I = T_G(\sigma \otimes \sigma')$ is trivial (so that η is also trivial) and therefore the multiplicity of η in $L(J)$ is simply $\dim L(J) = |J| = m_\sigma m_{\sigma'}$. In particular, the subspaces $W_{\sigma, \sigma'}^{j, j'}$ are mutually orthogonal and isomorphic to $\text{Ind}_{F \times F}^{F \wr C_2} (V_\sigma \otimes V_{\sigma'})$.

When $\sigma = \sigma'$, the inertia group of $V_\sigma \otimes V_{\sigma'}$ coincides with $F \wr C_2$ (indeed, in this case, $T_G(\sigma \otimes \sigma') \equiv T_G(\sigma \otimes \sigma) = G = C_2$). Therefore, the induction operation is trivial and we need only apply Corollary 3.3.12. We have $J = \{(i, j) : 1 \leq i, j \leq m_\sigma\}$ and the orbits of C_2 on J are $\{(i, j), (j, i)\}, 1 \leq i \neq j \leq m_\sigma$, and $\{(i, i)\}, i = 1, \ldots, m_\sigma$. This implies that $L(J)$ contains $\frac{1}{2} m_\sigma (m_\sigma + 1)$ times the trivial representation of C_2 (this corresponds to the case $\eta = \iota$, the trivial representation) and $\frac{1}{2} m_\sigma (m_\sigma - 1)$ times the nontrivial representation U (this corresponds to the case $\eta = \varepsilon$, the alternating representation). The subspace $W_{\sigma, +}^{i, j}$ corresponds to the choice $\eta = \iota$; in other words,

$W_{\sigma,+}^{i,j}$ is isomorphic to $V_\sigma \otimes V_\sigma$ as an $(F \wr C_2)$-representation. Analogously, for $i \neq j$ the subspace $W_{\sigma,-}^{i,j}$ corresponds to the choice $\eta = \varepsilon$, that is, $W_{\sigma,-}^{i,j}$ is isomorphic to $(V_\sigma \otimes V_\sigma) \otimes U$ as an $(F \wr C_2)$-representation.

The remaining part of the proof is now clear since the representations in (i) and (ii) exhaust the whole of $L(Y \times Y)$ (cf. the decomposition (3.47)).   □

3.3.3 The case when $L(Y)$ is multiplicity free

Suppose now that $L(Y)$ decomposes without multiplicity and that $L(Y) = \bigoplus_{i=0}^n V_i$ is the corresponding decomposition into inequivalent irreducible F-representations. We think of each V_i as a subspace of $L(Y)$; this means that if $v \in V_i$ then v is a function defined on Y and we denote by $v(y)$ its value on $y \in Y$. If σ_i is the representation of F on V_i then, for any $f \in F$, the unitary operator $\sigma_i(f) : V_i \to V_i$ is given by

$$[\sigma_i(f)v](y) = v(f^{-1}y) \qquad \text{for all } v \in V_i \text{ and } y \in Y. \tag{3.49}$$

For $u \in L(Z)$ and $g \in G$ we denote by gu the g-translate of u, that is, $gu(z) = u(g^{-1}z)$ for all $z \in Z$.

Denote by H the set of all maps $h : X \to \{0, 1, \dots, n\}$. If $h \in H$ and $v_x \in V_{h(x)}$ for all $x \in X$, we say that $\bigotimes_{x \in X} v_x \in L\left(Y^X\right)$ is a vector of *type h* in $L\left(Y^X\right)$. We denote by V_h the set of all vectors of type h in $L\left(Y^X\right)$, that is, $V_h = \bigotimes_{x \in X} V_{h(x)}$. Clearly we have the decomposition

$$L\left(Y^X\right) = \bigoplus_{h \in H} V_h$$

into irreducible F^X-representations.

In the present context, Lemma 3.3.4 has the following slightly more general form (note that $L(Y^X \times Z) \cong L(Y^X) \otimes L(Z)$ and we use Notation 3.3.3).

Lemma 3.3.15 *Let $(f, g) \in F \wr G$, $h \in H$, $\bigotimes_{x \in X} v_x \in V_h$ and $u \in L(Z)$. Then*

$$(f, g)\left[\left(\bigotimes_{x \in X} v_x\right) \otimes u\right] = \left(\bigotimes_{x \in X}\left[\sigma_{h(g^{-1}x)}(f(x))v_{g^{-1}x}\right]\right) \otimes gu.$$

Proof For all $(\varphi, z) \in Y^X \times Z$ we have

$$\left((f, g)\left[\left(\bigotimes_{x \in X} v_x\right) \otimes u\right]\right)(\varphi, z)$$

$$= \left[\left(\bigotimes_{x \in X} v_x\right) \otimes u\right]\left((f, g)^{-1}\varphi, g^{-1}z\right) \qquad \text{(by (3.34))}$$

$$= \left(\prod_{x \in X} v_x \left[f(gx)^{-1} \varphi(gx) \right] \right) (gu)(z) \qquad \text{(by (3.49), replacing } x \text{ by } g^{-1}x \text{)}$$

$$= \left(\prod_{x \in X} \left[\sigma_{h(g^{-1}x)}(f(x)) v_{g^{-1}x} \right] (\varphi(x)) \right) (gu)(z)$$

$$= \left(\left(\bigotimes_{x \in X} \left[\sigma_{h(g^{-1}x)}(f(x)) v_{g^{-1}x} \right] \right) \otimes gu \right) (\varphi, z).$$

$\square$

Fix $h \in H$ (equivalently, fix $\sigma \in \Sigma$ as in Section 3.3.1). Now $I = T_G(\sigma)$ coincides with the G-stabilizer of h (G acts on H in the obvious way: $(gh)(x) = h(g^{-1}x)$ for $h \in H$, $g \in G$ and $x \in X$). Let η be an irreducible I-representation contained in $L(Z)$ and denote by $U_1 \oplus U_2 \oplus \cdots \oplus U_m$ an orthogonal decomposition of the η-isotypic component in $L(Z)$. We regard each U_k as a subspace of $L(Z)$. Finally, let S be a system of representatives for the left cosets of I in G.

Theorem 3.3.16 *Set*

$$W_k = \bigoplus_{s \in S} (V_{sh} \otimes sU_k), \qquad k = 1, 2, \ldots, m.$$

Then we have that each W_k is isomorphic to the $(F \wr G)$-irreducible representation $\text{Ind}_{F \wr I}^{F \wr G} (\widetilde{\sigma} \otimes \overline{\eta})$ *and*

$$W_1 \oplus W_2 \oplus \cdots \oplus W_m$$

is an orthogonal decomposition of the $\text{Ind}_{F \wr I}^{F \wr G} (\widetilde{\sigma} \otimes \overline{\eta})$*-isotypic component of* $L\left(Y^X \times Z\right)$.

Proof In the present setting the space J (see (3.35)) is trivial since $m_i = 1$ for all $i = 1, 2, \ldots, n$, so that J reduces to the constant function with value 1 on X. As a consequence, (3.39) (after identifying $L(J \times Z)$ with $L(Z)$) becomes $\widehat{T}(v \otimes u) = v \otimes \mathcal{T}u$. Then from Theorem 3.3.8 we deduce that

$$\bigoplus_{k=1}^{m} \left(V_h \bigotimes U_k \right)$$

is an orthogonal decomposition of the $(\widetilde{\sigma} \otimes \overline{\eta})$-isotypic component in $L\left(Y^X \times Z\right)$. Now we apply (3.46). First note that $\{(1_F, s) : s \in S\}$ is a system of representatives for the left cosets of $F \wr I$ in $F \wr G$. If $s \in S$, $\bigotimes_{x \in X} v_x \in V_h$ and $u \in U_k$, Lemma 3.3.15 yields

$$(1_F, s)\left[\left(\bigotimes_{x\in X} v_x\right) \otimes u\right] = \left(\bigotimes_{x\in X} v_{s^{-1}x}\right) \otimes su.$$

In the setting of Theorem 1.2.27, this means that the irreducible $(F \wr G)$-representation in $L\left(Y^X \times Z\right)$ corresponding to the irreducible $(F \wr I)$-representation $V_h \otimes U_k$ is precisely $\bigoplus_{s \in S} (V_{sh} \otimes sU_k)$. $\qquad\square$

The following is a particular case of Corollary 3.3.11 but also a consequence of Theorem 3.3.16.

Corollary 3.3.17 *The multiplicity of* $\mathrm{Ind}_{F \wr I}^{F \wr G} (\widetilde{\sigma} \otimes \overline{\eta})$ *in* $L\left(Y^X \times Z\right)$ *is equal to the multiplicity of* η *in* $L(Z)$. $\qquad\square$

The representation-theoretic results in [66] are all particular cases of this example.

3.3.4 Exponentiation of finite Gelfand pairs

According to Corollary 3.3.17, when Z is trivial and $L(Y)$ is multiplicity free then $L\left(Y^X\right)$ is also multiplicity free. This may be translated into a result on the exponentiation of (finite) Gelfand pairs. We will analyze this fact more closely and with more elementary arguments; to this end, we will rearrange the notation. Let (F, H) be a finite Gelfand pair, let G be a finite group acting on a set X and consider the wreath product $F \wr G = F^X \rtimes G$. Set $Y = F/H$ and denote by $L(Y) = \oplus_{h=0}^n V_h$ the corresponding decomposition into spherical representations. Also, denote by $[n]^X$ the set consisting of all maps $\mathbf{i} : X \longrightarrow \{0, 1, 2, \ldots, n\}$ and set

$$V_{\mathbf{i}} = \bigotimes_{x\in X} V_{\mathbf{i}(x)}.$$

Then we have the decomposition

$$L\left(Y^X\right) = \bigoplus_{\mathbf{i}\in[n]^X} V_{\mathbf{i}}.$$

Denote by $\Gamma_0, \Gamma_1, \ldots, \Gamma_r$ the orbits of G on $[n]^X$ (with respect to the obvious action defined by setting $g\mathbf{i}(x) = \mathbf{i}(g^{-1}x)$, for all $g \in G$ and $x \in X$). Set

$$W_j = \bigoplus_{\mathbf{i}\in\Gamma_j} V_{\mathbf{i}}.$$

Finally, denoting by ϕ_i the spherical function in V_i, $i = 0, 1, 2, \ldots, n$, we set

$$\Phi_j = \frac{1}{|\Gamma_j|}\sum_{\mathbf{i}\in\Gamma_j}\bigotimes_{x\in X}\phi_{\mathbf{i}(x)}$$

for $j = 0, 1, 2, \ldots, r$.

Theorem 3.3.18 *With all the assumptions above, we have:*

(i) $(F \wr G)/(H \wr G) \cong Y^X$;

(ii) $(F \wr G, H \wr G)$ *is a Gelfand pair;*

(iii) $L(X^Y) = \bigoplus_{j=0}^{r} W_j$ *is the decomposition of $L(Y^X)$ into spherical representations;*

(iv) Φ_j *is the spherical function belonging to W_j for all $j = 0, 1, 2, \ldots, r$.*

$\square$

Particular cases of this construction have been studied recently (mostly from the point of view of the theory of special functions) by Mizukawa [57] and Akazawa and Mizukawa [1]. See also Mizukawa and Tanaka [58].

Exercise 3.3.19 Prove the claims in Theorem 3.3.18. Also prove that W_j is isomorphic to $\mathrm{Ind}_{F \wr I_j}^{F \wr G} V_i$, where $\mathbf{i} \in \Gamma_j$ and I_j is the G-stabilizer of $\mathbf{i}$.

Exercise 3.3.20 Deduce the results of Section 5.4 ("The group theoretical approach to the Hamming scheme") in [12] as a particular case of Theorem 3.3.18.

3.4 Harmonic analysis on finite lamplighter spaces

Here we consider the case $F \equiv Y = C_2$ in the setting of Section 3.3.3 (but we use additive notation, thus identifying C_2 with $\{0, 1\}$). However, rather than applying the general theory previously developed, it is now worthwhile to use the results in Section 2.5 on the representation theory of groups of the form $C_2 \wr G$ to develop a more direct and elementary approach.

3.4.1 Finite lamplighter spaces

First we introduce the specific notation that we shall use in the present section. Again, G is a finite group and X and Z are homogeneous G-spaces. We fix $z_0 \in Z$ and denote by $H = \{g \in G : gz_0 = z_0\}$ its stabilizer, so that as G-spaces Z and G/H may be identified. We consider the wreath product or *(finite) lamplighter group* $C_2 \wr G = C_2^X \rtimes G$; we will use the notation in Sections 2.3.2 and 2.5. Now the action of the group $C_2 \wr G$ on $C_2^X \times Z$ may be described by setting

$$(\omega, g)(\theta, z) = ((\omega, g)\theta, gz), \qquad \text{where } (\omega, g)\theta = \omega + g\theta$$

for all $(\omega, g) \in C_2 \wr G$, $\theta \in C_2^X$ and $z \in Z$. The stabilizer of $(0_{C_2}, z_0)$ is just the subgroup $\widetilde{H} = \{(0_{C_2}, h) : h \in H\} \cong H$, so that $C_2^X \times Z = (C_2 \wr G)/\widetilde{H}$ as $(C_2 \wr G)$-spaces.

Definition 3.4.1 A $(C_2 \wr G)$-homogeneous space of the form $C_2^X \times Z$, as above, will be called a *(finite) lamplighter space*.

We shall use the isomorphism

$$L\left(C_2^X \times Z\right) \cong L\left(C_2^X\right) \otimes L(Z)$$

(see (3.3)). Now Lemma 3.3.15 has a more specific form; we write it as a formula for the action of an element of $C_2 \wr G$ on a tensor product of the kind $\chi_\theta \otimes f$ (where $\chi_\theta \in \widehat{C_2^X}$, see Section 2.5).

Proposition 3.4.2 *Let* $(\omega, g) \in C_2 \wr G$, $\theta \in C_2^X$ *and* $f \in L(Z)$. *Then*

$$(\omega, g)(\chi_\theta \otimes f) = \chi_{g\theta}(\omega) [\chi_{g\theta} \otimes gf].$$

Proof Given $(\sigma, z) \in C_2^X \times Z$, we have

$$
\begin{aligned}
[(\omega, g)(\chi_\theta \otimes f)](\sigma, z) &= (\chi_\theta \otimes f)[(\omega, g)^{-1}(\sigma, z)] \\
&= (\chi_\theta \otimes f)(g^{-1}\omega + g^{-1}\sigma, g^{-1}z) \\
&= \chi_\theta(g^{-1}\omega + g^{-1}\sigma) f(g^{-1}z) \\
&= \chi_{g\theta}(\omega) [\chi_{g\theta} \otimes gf](\sigma, z).
\end{aligned}
$$

$\square$

In the notation of Theorem 2.5.1, for any $\theta \in \Theta$ (a fixed system of representatives of the G-orbits on C_2^X) we choose a system S_θ of representatives for the left cosets of G_θ (the G-stabilizer of χ_θ) in G (with $1_G \in S_\theta$), so that $G = \coprod_{s \in S_\theta} sG_\theta$.

Let $\theta \in \Theta$ and let V be a G_θ-invariant and irreducible subspace of $L(Z)$; we then denote by η the corresponding representation in $\widehat{G_\theta}$, but for $f \in V$ and $g \in G$ we simply write gf for the g-translate of f. The following result is an immediate consequence of Proposition 3.4.2.

Corollary 3.4.3 *Let* $(\omega, g) \in C_2 \wr G$, $s \in S_\theta$ *and* $f \in sV$. *Suppose that* $gs = th$ *with* $h \in G_\theta$ *and* $t \in S_\theta$. *Then*

$$(\omega, g)(\chi_{s\theta} \otimes f) = \chi_{t\theta} \otimes f',$$

where $f' = \chi_{t\theta}(\omega) ths^{-1}f \in tV$.

Lemma 3.4.4 *Let* $\theta, \theta' \in \Theta$ *and* $s \in S_\theta$ *(resp.* $s' \in S_{\theta'}$*), and let* V *(resp.* V'*) be a* G_θ-*invariant (resp.* $G_{\theta'}$-*invariant) subspace in* $L(Z)$. *Then, for* $f \in V$ *and* $f' \in V'$, *we have*

$$\langle \chi_{s\theta} \otimes sf, \chi_{s'\theta'} \otimes s'f' \rangle_{L(C_2^X \times Z)} = \delta_{\theta,\theta'} \delta_{s,s'} 2^{|X|} \langle f, f' \rangle_{L(Z)}.$$

Proof We have $s\theta = s'\theta'$ if and only if $\theta = \theta'$ and $s = s'$. Therefore

$$\langle \chi_{s\theta} \otimes sf, \chi_{s'\theta'} \otimes s'f' \rangle_{L(C_2^X \times Z)} = \langle \chi_{s\theta}, \chi_{s'\theta'} \rangle_{L(C_2^X)} \langle sf, s'f' \rangle_{L(Z)}$$

$$= \delta_{\theta,\theta'} \delta_{s,s'} 2^{|X|} \langle f, f' \rangle_{L(Z)}.$$

$\square$

Lemma 3.4.5 *We have the equivalence*

$$\mathrm{Ind}_{C_2 \wr G_\theta}^{C_2 \wr G} (\widetilde{\chi}_\theta \otimes \overline{\eta}) \sim \bigoplus_{s \in S_\theta} \{ \chi_{s\theta} \otimes sf : f \in V \}. \tag{3.50}$$

Proof From Corollary 3.4.3 (or Proposition 3.4.2) we deduce that, for $(\omega, h) \in C_2 \wr G_\theta$ and $f \in V$,

$$(\omega, h)(\chi_\theta \otimes f) = \chi_\theta \otimes [\chi_\theta(\omega)hf],$$

that is, the subspace $\{ \chi_\theta \otimes f : f \in V \}$ of $L(C_2^X \times Z)$ is $(C_2 \wr G_\theta)$-invariant and the corresponding $(C_2 \wr G_\theta)$-representation is equivalent to $\widetilde{\chi}_\theta \otimes \overline{\eta}$: recall (2.46). Analogously, for $s \in S_\theta$ we have

$$(\mathbf{0}_{C_2}, s)(\chi_\theta \otimes f) = \chi_{s\theta} \otimes sf.$$

This implies that

$$\bigoplus_{s \in S_\theta} \{ \chi_{s\theta} \otimes sf : f \in V \} \equiv \bigoplus_{s \in S_\theta} (\mathbf{0}_{C_2}, s) \{ \chi_\theta \otimes f : f \in V \}. \tag{3.51}$$

Moreover, the space (3.51) is $(C_2 \wr G)$-invariant (apply Corollary 3.4.3) and from Lemma 3.4.4 it follows that it is an orthogonal direct sum. In this way, we can apply Proposition 1.1.9 (the set $\{ (\mathbf{0}_{C_2}, s) : s \in S_\theta \}$ is a system of representatives for the left cosets of $C_2 \wr G_\theta$ in $C_2 \wr G$) and (3.50) is proved. $\square$

For each $\theta \in \Theta$, let

$$L(Z) = \bigoplus_{i=0}^{n(\theta)} m_{\theta,i} V_{\theta,i} \tag{3.52}$$

be the isotypic decomposition of $L(Z)$ into irreducible G_θ-representations. This means that, for different values of i, the corresponding representations are inequivalent; $m_{\theta,i}$ is the multiplicity of $V_{\theta,i}$ in $L(Z)$. Moreover, let

$$m_{\theta,i} V_{\theta,i} = V_{\theta,i}^1 \oplus V_{\theta,i}^2 \oplus \cdots \oplus V_{\theta,i}^{m_{\theta,i}} \tag{3.53}$$

be an explicit orthogonal decomposition of the isotypic component $m_{\theta,i} V_{\theta,i}$, (so that each $V_{\theta,i}^j$ is equivalent to $V_{\theta,i}$). With each $V_{\theta,i}^j$ we associate the space

$$W_{\theta,i}^j = \bigoplus_{s \in S_\theta} \left(\chi_{s\theta} \otimes sf : f \in V_{\theta,i}^j \right) \tag{3.54}$$

(see Lemma 3.4.5) and we set

$$m_{\theta,i} W_{\theta,i} = W_{\theta,i}^1 \oplus W_{\theta,i}^2 \oplus \cdots W_{\theta,i}^{m_{\theta,i}}. \tag{3.55}$$

Theorem 3.4.6 *The decomposition of $L\left(C_2^X \times Z\right)$ into irreducible $(C_2 \wr G)$-representations is given by*

$$L\left(C_2^X \times Z\right) = \bigoplus_{\theta \in \Theta} \bigoplus_{i=0}^{n(\theta)} m_{\theta,i} W_{\theta,i}, \tag{3.56}$$

where each $m_{\theta,i} W_{\theta,i}$ is an isotypic component with (3.55) an explicit orthogonal decomposition into irreducible (equivalent) representations.

Proof From Theorem 2.5.1 and Lemma 3.4.5 it follows that the representations $W_{\theta,i}^1, W_{\theta,i}^2, \ldots, W_{\theta,i}^{m_{\theta,i}}$ are irreducible and equivalent. From Lemma 3.4.4 it follows that the right-hand side of (3.56) (resp. of (3.55)) is an orthogonal direct sum. We end the proof by showing, by considering dimensions, that the direct sum on the right-hand side of (3.56) fills the whole of $L\left(C_2^X \times Z\right)$. This is easy:

$$\sum_{\theta \in \Theta} \sum_{i=0}^{n(\theta)} m_{\theta,i} \dim W_{\theta,i} = \sum_{\theta \in \Theta} \sum_{i=0}^{n(\theta)} |S_\theta| m_{\theta,i} \dim V_{\theta,i}$$

$$= \sum_{\theta \in \Theta} \left|\frac{G}{G_\theta}\right| \dim L(Z)$$

$$= |C_2^X| |Z|$$

$$= \dim L(C_2^X \times Z).$$

$\square$

We can now state a particular case of Corollary 3.3.17.

Corollary 3.4.7 *We have that the multiplicity of the irreducible representation $\mathrm{Ind}_{C_2 \wr G_\theta}^{C_2 \wr G} \left(\widetilde{\chi}_\theta \otimes \overline{\eta}\right)$ in $L\left(C_2^X \times Z\right)$ equals the multiplicity of η in the decomposition of $L(Z)$ under the action of G_θ.* $\square$

3.4.2 Spectral analysis of an invariant operator

We now examine the spectral analysis of a linear self-adjoint invariant operator $\mathcal{M}$ on a finite lamplighter space. By virtue of Proposition 1.2.23, there exist decompositions of the form (3.55) such that each $W_{\theta,i}^j$ is an eigenspace of $\mathcal{M}$. The next theorem reduces the spectral analysis of $\mathcal{M}$ to the spectral analysis of a collection $\{M_\theta : \theta \in \Theta\}$ of G_θ-invariant operators on $L(Z)$.

Theorem 3.4.8   *Let* $\mathcal{M} : L\left(C_2^X \times Z\right) \rightarrow L\left(C_2^X \times Z\right)$ *be a linear, self-adjoint,* $(C_2 \wr G)$-*invariant operator. Then the following hold:*

(i) *For each* $\theta \in \Theta$, *there exists a* G_θ-*invariant, linear, self-adjoint operator* $M_\theta : L(Z) \rightarrow L(Z)$ *such that*

$$\mathcal{M}(\chi_\theta \otimes f) = \chi_\theta \otimes M_\theta f \qquad (3.57)$$

for all $f \in L(Z)$.

(ii) *If* $V_{\theta,i}^j$ *in (3.53) is an eigenspace of* M_θ *with eigenvalue* $\lambda_{\theta,i}^j$ *then the corresponding space* $W_{\theta,i}^j$ *in (3.56) is an eigenspace of* $\mathcal{M}$, *with the same eigenvalue* $\lambda_{\theta,i}^j$.

Proof   (i) From Proposition 3.4.2 and the $(C_2 \wr G)$-invariance of $\mathcal{M}$, it follows that

$$(\omega, g)\mathcal{M}(\chi_\theta \otimes f) = \chi_{g\theta}(\omega)\mathcal{M}(\chi_{g\theta} \otimes gf) \qquad (3.58)$$

for all $(\omega, g) \in C_2 \wr G$. In particular, when $g = 1_G$, (3.58) gives

$$(\omega, 1_G)\mathcal{M}(\chi_\theta \otimes f) = \chi_\theta(\omega)\mathcal{M}(\chi_\theta \otimes f).$$

This implies that $\mathcal{M}(\chi_\theta \otimes f)$ belongs to $\{\chi_\theta \otimes f' : f' \in L(Z)\}$, the χ_θ-isotypic component in the decomposition of $L\left(C_2^X \times Z\right)$ under the action of C_2^X. Therefore, for any $f \in L(Z)$ there exists $f' \in L(Z)$ such that $\mathcal{M}(\chi_\theta \otimes f) = \chi_\theta \otimes f'$. Setting $M_\theta f = f'$, we define a linear self-adjoint operator $M_\theta : L(Z) \rightarrow L(Z)$ satisfying (3.57).

Moreover, we have

$$\begin{aligned}
\chi_{g\theta} \otimes gM_\theta f &= (\mathbf{0}_{C_2}, g)(\chi_\theta \otimes M_\theta f) && \text{(by Proposition 3.4.2)} \\
&= (\mathbf{0}_{C_2}, g)\mathcal{M}(\chi_\theta \otimes f) && \text{(by (3.57))} \\
&= \mathcal{M}(\chi_{g\theta} \otimes gf) && \text{(by (3.58))} \\
&= \chi_{g\theta} \otimes M_{g\theta}(gf) && \text{(again by (3.57)).}
\end{aligned}$$

Therefore

$$gM_\theta f = M_{g\theta}(gf) \qquad (3.59)$$

for all $g \in G$ and $f \in L(Z)$. In particular, when $g \in G_\theta$ equation (3.59) becomes $gM_\theta f = M_\theta(gf)$, showing the G_θ-invariance of M_θ.

(ii) Now let $V_{\theta,i}^j$ (see (3.53)) be an eigenspace of M_θ, with corresponding eigenvalue $\lambda_{\theta,i}^j$, that is, $M_\theta f = \lambda_{\theta,i}^j f$ for all $f \in V_{\theta,i}^j$. Then for all $s \in S_\theta$ we have

$$\mathcal{M}(\chi_{s\theta} \otimes sf) = \chi_{s\theta} \otimes M_{s\theta}(sf) \qquad \text{(by (3.57))}$$
$$= \chi_{s\theta} \otimes sM_\theta f \qquad \text{(by (3.59))}$$
$$= \lambda^j_{\theta,i}(\chi_{s\theta} \otimes sf).$$

This means that $W^j_{\theta,i}$ (cf. (3.56)) is an eigenspace of $\mathcal{M}$, with the same eigenvalue $\lambda^j_{\theta,i}$. □

In other words, the action of the group $C_2 \wr G$ collects together all the $\mathcal{M}$-eigenspaces $\{\chi_{s\theta} \otimes sf : f \in V^j_{\theta,i}\}$ into an irreducible representation.

3.4.3 Spectral analysis of lamplighter graphs

The theory developed in Section 3.4.2 has a graph-theoretic analogue that uses only invariance under the action of C_2^X. Let (X, E) be a finite, simple, undirected graph without loops (X is the vertex set and E, a collection of 2-subsets of X, is the edge set). We write $x \sim y$ to denote that two distinct vertices $x, y \in X$ are connected (or adjacent), that is, $\{x, y\} \in E$.

Definition 3.4.9 The *lamplighter graph* associated with (X, E) is the finite graph (X', E'), where

$$X' = \{0, 1\}^X \times X = \Big\{(\omega, x) : \omega \in \{0, 1\}^X, x \in X\Big\},$$

and two vertices $(\omega, x), (\theta, y) \in X'$ are connected if $x \sim y$ (in X) and $\omega(z) = \theta(z)$ for all $z \neq x, y$.

The associated *adjacency operator* $\mathcal{A}_{X'} : L(X') \to L(X')$ is the linear operator defined by setting

$$(\mathcal{A}_X \mathcal{F})(\omega, x) = \sum_{(\theta, y) \sim (\omega, x)} \mathcal{F}(\theta, y)$$

for all $\mathcal{F} \in L(X')$ and $(\omega, x) \in X'$. Since $L(X') \equiv L\left(\{0, 1\}^X\right) \otimes L(X)$, it is useful to write down the action of $\mathcal{A}_{X'}$ on an elementary tensor:

$$[\mathcal{A}_X(F \otimes f)](\omega, x)$$
$$= \sum_{y \sim x}[F(\omega) + F(\omega + \delta_x) + F(\omega + \delta_y) + F(\omega + \delta_x + \delta_y)]f(y) \quad (3.60)$$

for all $F \in L\left(\{0, 1\}^X\right)$, $f \in L(X)$ and $(\omega, x) \in \{0, 1\}^X \times X$.

For $\theta \in \{0, 1\}^X$, we set $X_\theta = \{x \in X : \theta(x) = 0\}$ and $E_\theta = \{\{x, y\} \in E : x, y \in X_\theta\}$. In this way, (X_θ, E_θ) is a subgraph of X. We define a linear operator $A_\theta : L(X) \to L(X)$ by setting

$$(A_\theta f)(x) = \begin{cases} \displaystyle\sum_{\substack{y \in X_\theta: \\ y \sim x}} f(y) & \text{if } \theta(x) = 0 \\ 0 & \text{if } \theta(x) = 1 \end{cases}$$

for all $f \in L(X)$ and $x \in X$. In other words, A_θ is the adjacency operator of (X_θ, E_θ) trivially extended to the whole of $L(X)$.

The following provides a graph-theoretic analogue of (3.57).

Lemma 3.4.10   *For all $f \in L(X)$ and $\theta \in \{0, 1\}^X$ we have*

$$A_{X'}(\chi_\theta \otimes f) = \chi_\theta \otimes 4 A_\theta f.$$

Proof   Applying (3.60) we get

$$[A_{X'}(\chi_\theta \otimes f)](x, \omega)$$
$$= \sum_{y \sim x} [\chi_\theta(\omega) + \chi_\theta(\omega + \delta_x) + \chi_\theta(\omega + \delta_y) + \chi_\theta(\omega + \delta_x + \delta_y)] f(y).$$

$$(3.61)$$

Since

$$\chi_\theta(\omega) + \chi_\theta(\omega + \delta_x) + \chi_\theta(\omega + \delta_y) + \chi_\theta(\omega + \delta_x + \delta_y)$$
$$= \chi_\theta(\omega)[1 + \chi_\theta(\delta_x) + \chi_\theta(\delta_y) + \chi_\theta(\delta_x + \delta_y)]$$
$$= \chi_\theta(\omega)[1 + (-1)^{\theta(x)} + (-1)^{\theta(y)} + (-1)^{\theta(x)+\theta(y)}]$$
$$= \begin{cases} 4\chi_\theta(\omega) & \text{if } \theta(x) = \theta(y) = 0 \\ 0 & \text{otherwise,} \end{cases}$$

it follows that (3.61) gives

$$[A_{X'}(\chi_\theta \otimes f)](x, \omega) = 4\chi_\theta(\omega) \sum_{\substack{y \in X_\theta: \\ y \sim x}} f(y) \equiv \chi_\theta(\omega) \otimes 4 A_\theta f$$

if $\theta(x) = 0$ and

$$[A_X(\chi_\theta \otimes f)](x, \omega) = 0 \equiv \chi_\theta(\omega) \otimes 4 A_\theta f$$

otherwise. $\square$

In the following, we will show that Lemma 3.4.10 enables us to express the spectrum of X' in terms of the spectra of the subgraphs of (X_θ, E_θ), $\theta \in \{0, 1\}^X$. Setting $V_\theta = \{\chi_\theta \otimes f : f \in L(X)\}$, we have the orthogonal decomposition

$$L(X') = \bigoplus_{\theta \in \{0,1\}^X} V_\theta. \qquad (3.62)$$

This is the C_2^X-isotypic decomposition of $L(X')$.

Proposition 3.4.11 *Define the linear operator* $\widetilde{Q}_\theta : L(X') \rightarrow L(X)$ *by setting*

$$(\widetilde{Q}_\theta F)(x) = \frac{1}{2^{|X|}} \sum_{\omega \in \{0,1\}^X} F(\omega, x) \chi_\theta(\omega)$$

for all $F \in L(X')$. *Then we have the orthogonal decomposition*

$$F = \sum_{\theta \in \{0,1\}^X} \chi_\theta \otimes \widetilde{Q}_\theta F. \tag{3.63}$$

In particular, the orthogonal projection $Q_\theta : L(X') \rightarrow V_\theta$ *satisfies*

$$Q_\theta F = \chi_\theta \otimes \widetilde{Q}_\theta F.$$

Proof First we recall the orthogonality relations in $\{0,1\}^X$. For $\omega, \xi \in \{0,1\}^X$ we have

$$\frac{1}{2^{|X|}} \sum_{\theta \in \{0,1\}^X} \chi_\theta(\omega) \chi_\theta(\xi) \equiv \frac{1}{2^{|X|}} \sum_{\theta \in \{0,1\}^X} \chi_\omega(\theta) \chi_\xi(\theta) = \begin{cases} 1 & \text{if } \omega = \xi \\ 0 & \text{if } \omega \neq \xi. \end{cases}$$

Indeed, if $\omega = \xi$ then $\chi_\omega(\theta) \chi_\xi(\theta) = 1$ for all $\omega \in \{0,1\}^X$. Moreover, if $\omega \neq \xi$ then there exists $\eta \in \{0,1\}$ such that $\chi_\omega(\eta) \chi_\xi(\eta) = -1$, and thus

$$- \sum_{\theta \in \{0,1\}^X} \chi_\omega(\theta) \chi_\xi(\theta) = \chi_\omega(\eta) \chi_\xi(\eta) \sum_{\theta \in \{0,1\}^X} \chi_\omega(\theta) \chi_\xi(\theta)$$

$$= \sum_{\theta \in \{0,1\}^X} \chi_\omega(\theta + \eta) \chi_\xi(\theta + \eta)$$

$$= \sum_{\theta \in \{0,1\}^X} \chi_\omega(\theta) \chi_\xi(\theta),$$

which forces $\sum_{\theta \in \{0,1\}^X} \chi_\omega(\theta) \chi_\xi(\theta) = 0$. Therefore, for $(\omega, x) \in \{0,1\}^X \times X$ we have

$$\left(\sum_{\theta \in \{0,1\}^X} \chi_\theta \otimes \widetilde{Q}_\theta F \right)(\omega, x) = \sum_{\theta \in \{0,1\}^X} \chi_\theta(\omega) \left(\widetilde{Q}_\theta F \right)(x)$$

$$= \sum_{\xi \in \{0,1\}^X} F(\xi, x) \frac{1}{2^{|X|}} \sum_{\theta \in \{0,1\}^X} \chi_\theta(\omega) \chi_\theta(\xi)$$

$$= F(\omega, x).$$

This proves (3.63), and it is easy to check that it is an orthogonal decomposition. $\qquad\square$

Now let

$$A_\theta = \lambda_{\theta,1} P_{\theta,1} + \lambda_{\theta,2} P_{\theta,2} + \cdots + \lambda_{\theta,h(\theta)} P_{\theta,h(\theta)} \qquad (3.64)$$

denote the spectral decomposition of A_θ. That is, $\lambda_{\theta,1}, \lambda_{\theta,2}, \ldots, \lambda_{\theta,h(\theta)}$ are the distinct nonzero eigenvalues and $P_{\theta,j}$ is the orthogonal projection of $L(X)$ onto the eigenspace of $\lambda_{\theta,j}$. If $X_\theta \subsetneq X$ then A_θ admits $L(X \setminus X_\theta)$ as an eigenspace with eigenvalue equal to zero; this is omitted in (3.64).

Theorem 3.4.12 *The operator $A_{X'}$ has the following spectral decomposition:*

$$A_{X'} = \sum_{\theta \in \{0,1\}^X} \sum_{j=1}^{h(\theta)} 4\lambda_{\theta,j} \left(\chi_\theta \otimes P_{\theta,j} \widetilde{Q}_\theta\right),$$

where

$$(\chi_\theta \otimes P_{\theta,j} \widetilde{Q}_\theta)F = \chi_\theta \otimes P_{\theta,j} \widetilde{Q}_\theta F$$

for all $F \in L(X')$. The zero eigenvalues (in particular those corresponding to the space $L(X \setminus X_\theta)$) are omitted, and the eigenvalues $\lambda_{\theta,j}$, $\theta \in \{0, 1\}^X$, $j = 1, 2, \ldots, h(\theta)$, are not necessarily distinct.

Proof   The proof follows immediately from (3.63), Lemma 3.4.10 and (3.64). Indeed, we have

$$\begin{aligned}
A_{X'}F &= A_{X'}\left(\sum_{\theta \in \{0,1\}^X} \chi_\theta \otimes \widetilde{Q}_\theta F\right) \\
&= \sum_{\theta \in \{0,1\}^X} \chi_\theta \otimes 4A_\theta \widetilde{Q}_\theta F \\
&= \sum_{\theta \in \{0,1\}^X} \sum_{j=1}^{h(\theta)} 4\lambda_{\theta,j} \left(\chi_\theta \otimes P_{\theta,j} \widetilde{Q}_\theta F\right)
\end{aligned}$$

for all $F \in L(X')$.   □

Corollary 3.4.13 *If $V_{\theta,j}$ is the eigenspace of A_θ corresponding to the eigenvalue $\lambda_{\theta,j}$, $j = 0, 1, \ldots, h(\theta)$ (with $V_{\theta,0}$ the eigenspace of $\lambda_{\theta,0} = 0$), then $W_{\theta,j} = \{\chi_\theta \otimes f : f \in V_{\theta,j}\}$ is the eigenspace of A_θ corresponding to $4\lambda_{\theta,j}$.*

3.4.4 The lamplighter on the complete graph

In this subsection we examine the case $G = S_n$, $X \equiv Z = \{1, 2, \ldots, n\}$, with the natural action. We use the results in Section 3.4.1 for the representation-theoretic aspects and those in Section 3.4.3 for the spectral analysis of the related graph. Using the standard notation in the representation theory of the symmetric group (see Section 3.2.1), if A is a finite set of cardinality $|A| = n$ then we denote by $S^n(A)$ the set of all constant complex-valued functions defined on A and by $S^{n-1,1}(A)$ the set of all complex-valued functions f defined on A such that $\sum_{a \in A} f(a) = 0$. We recall the following elementary fact:

$$L(X) = S^n(X) \oplus S^{n-1,1}(X) \tag{3.65}$$

is the decomposition of $L(X)$ into irreducible S_n-representations (cf. Exercise 1.2.34). For a subgroup of S_n of the form $S_k \times S_{n-k}$, $0 \leq k \leq n$, we denote by B_k the k-subset of X fixed by $S_k \times S_{n-k}$. In particular, the orbits of $S_k \times S_{n-k}$ on X are B_k and $X \setminus B_k$.

Theorem 3.4.14 *Set*

$$W^1_{k;0} = \langle \chi_\theta \otimes f : |X_\theta| = k, \ f|_{X_\theta} \in S^{(k)}(X_\theta) \ and \ f|_{X \setminus X_\theta} \equiv 0 \rangle,$$
$$W^0_{k;0} = \langle \chi_\theta \otimes f : |X_\theta| = k, \ f|_{X_\theta} \equiv 0 \ and \ f|_{X \setminus X_\theta} \in S^{(n-k)}(X \setminus X_\theta) \rangle,$$
$$W_{k;1} = \langle \chi_\theta \otimes f : |X_\theta| = k, \ f|_{X_\theta} \in S^{k-1,1}(X_\theta) \ and \ f|_{X \setminus X_\theta} \equiv 0 \rangle,$$
$$W_{k;2} = \langle \chi_\theta \otimes f : |X_\theta| = k, \ f|_{X_\theta} \equiv 0 \ and \ f|_{X \setminus X_\theta} \in S^{n-k-1,1}(X \setminus X_\theta) \rangle.$$

Then, a decomposition of the permutation representation of $C_2 \wr S_n$ on $C_2^X \times X$ is given by

$$L(C_2^X \times X)$$
$$= \left(W^0_{0;0} \oplus W_{0;2} \right) \oplus \left[\bigoplus_{k=1}^{n-1} \left(W^0_{k;0} \oplus W^1_{k;0} \oplus W_{k;1} \oplus W_{k;2} \right) \right] \oplus \left(W^1_{n;0} \oplus W_{n;1} \right).$$

Proof By using (3.65), we can decompose the space $L(X)$ into irreducible $(S_k \times S_{n-k})$-representations for $1 \leq k \leq n-1$:

$$L(X) = L(B_k) \oplus L(X \setminus B_k)$$
$$= S^k(B_k) \oplus S^{k-1,1}(B_k) \oplus S^{n-k}(X \setminus B_k) \oplus S^{n-k-1,1}(X \setminus B_k).$$

An application of Theorem 3.4.6 ends the proof. $\square$

In the notation of Section 2.5.2 the representations of $C_2 \wr S_n$ on $W^1_{k;0}$ and $W^1_{k;0}$ are both isomorphic to $\rho_{[(k,0);(n-k,0)]}$, the representation on $W_{k;1}$ is isomorphic to $\rho_{[(k-1,1);(n-k,0)]}$ and the representation on $W_{k;2}$ is isomorphic to $\rho_{[(k,0);(n-k-1,1)]}$. In particular, the representations $W_{k;1}$ and $W_{k;2}$ have multiplicity 1, while the representations $W_{k;0}$ have multiplicity 2.

Now let (X, E) be the *complete graph* on n vertices and identify X with $\{1, 2, \ldots, n\}$; thus $E = X \times X \setminus \{(x, x) : x \in X\}$. The eigenspaces of the adjacency operator on the complete graph on n vertices are $S^n(X)$ and $S^{n-1,1}(X)$, with corresponding eigenvalues $n - 1$ and -1, respectively. Moreover, for $\theta \in \{0, 1\}^X$ the graph (X_θ, E_θ) is the complete graph on $|X_\theta|$ vertices.

For $\theta \in \{0, 1\}^X$, we consider the linear projection $P_\theta : L(X) \to L(X)$ defined by setting

$$
P_\theta f(x) = \begin{cases} \frac{1}{|X_\theta|} \sum_{y \in X_\theta} f(y) & \text{if } x \in X_\theta \\ 0 & \text{if } x \notin X_\theta \end{cases}
$$

for all $f \in L(X)$. Then, for $|X_\theta| > 1$, the spectral decomposition of the operator A_θ is given by

$$
A_\theta = (|X_\theta| - 1)P_\theta - (R_\theta - P_\theta),
$$

where $R_\theta : L(X) \to L(X_\theta)$ is the orthogonal projection from $L(X)$ onto $L(X_\theta)$. In the notation of Theorem 3.4.12 we have $h(\theta) = 2, \lambda_{\theta,1} = (|X_\theta|-1)$, $\lambda_{\theta,2} = -1, P_{\theta,1} = P_\theta$ and $P_{\theta,2} = R_\theta - P_\theta$. Clearly, if $|X_\theta| = 1$ then $X_\theta = \{x\}$ for some $x \in X$ and $A_\theta \equiv 0$. Moreover, in the notation of Theorem 3.4.14 we have the following:

$$
W_{k;0}^1 = \bigoplus_{\substack{\theta \in C_2^X \\ |X_\theta|=k}} \text{Ran}(P_\theta)
$$

is the eigenspace with eigenvalue $4(k - 1)$ for $k = 1, 2, \ldots, n$;

$$
W_{k;1} = \bigoplus_{\substack{\theta \in C_2^X: \\ |X_\theta|=k}} \text{Ran}(R_\theta - P_\theta)
$$

for $k = 1, 2, \ldots, n$, and

$$
\bigoplus_{k=1}^n W_{k;1} = \bigoplus_{k=1}^n \left[\bigoplus_{\substack{\theta \in C_2^X: \\ |X_\theta|=k}} \text{Ran}(R_\theta - P_\theta) \right]
$$

is the eigenspace with eigenvalue -4; finally,

$$\left(\bigoplus_{k=0}^{n-1} W_{k;0}^0\right) \oplus \left(\bigoplus_{k=1}^{n-1} W_{k;2}\right)$$

is the eigenspace with eigenvalue 0. Note that the operator $\mathcal{A}_{X'}$ is not in the center of the commutant algebra (see Propositions 1.2.23 and 1.2.25). Indeed, $W_{k;0}^0$ and $W_{k;0}^1$ are equivalent but correspond to different eigenvalues, namely 0 and $4(k-1)$ respectively.

References

[1] H. Akazawa and H. Mizukawa, Orthogonal polynomials arising from the wreath products of a dihedral group with a symmetric group. *J. Combin. Theory Ser. A* **104** (2003), no. 2, 371–380.

[2] J.L. Alperin and R.B. Bell, *Groups and Representations*. Graduate Texts in Mathematics, vol. 162, Springer-Verlag, New York, 1995.

[3] R.A. Bailey, C.E. Praeger, C.A. Rowley and T.P. Speed, Generalized wreath products of permutation groups. *Proc. London Math. Soc.* (3) **47** (1983), no. 1, 69–82.

[4] E. Bannai and T. Ito, *Algebraic Combinatorics*, Benjamin, 1984.

[5] Ya.G. Berkovich and E.M. Zhmud, *Characters of Finite Groups. Part 1*. Translations of Mathematical Monographs, vol. 172, American Mathematical Society, 1998.

[6] Ya.G. Berkovich and E.M. Zhmud, *Characters of Finite Groups. Part 2*. Translated from the Russian manuscript by P. Shumyatsky, V. Zobina and Ya.G. Berkovich. Translations of Mathematical Monographs, vol. 181. American Mathematical Society, 1999.

[7] D. Bump, *Lie Groups*. Graduate Texts in Mathematics, vol. 225, Springer-Verlag, 2004.

[8] T. Ceccherini-Silberstein, Yu. Leonov, F. Scarabotti and F. Tolli, Generalized Kaloujnine groups, uniseriality and height of automorphisms. *Internat. J. Algebra Comput.* **15** (2005), no. 3, 503–527.

[9] T. Ceccherini-Silberstein, F. Scarabotti and F. Tolli, Trees, wreath products and finite Gelfand pairs. *Adv. Math.* **206** (2006), no. 2, 503–537.

[10] T. Ceccherini-Silberstein, F. Scarabotti and F. Tolli, Finite Gelfand pairs and applications to probability and statistics. *J. Math. Sci.* **141** (2007), no. 2, 1182–1229.

[11] T. Ceccherini-Silberstein, F. Scarabotti and F. Tolli, *Harmonic Analysis on Finite Groups: Representation Theory, Gelfand Pairs and Markov Chains*. Cambridge Studies in Advanced Mathematics, vol. 108, Cambridge University Press, 2008.

[12] T. Ceccherini-Silberstein, A. Machì, F. Scarabotti and F. Tolli, Induced representation and Mackey theory. *J. Math. Sci.* **156** (2009), no. 1, 11–28.

[13] T. Ceccherini-Silberstein, F. Scarabotti and F. Tolli, Clifford theory and applications. *J. Math. Sci.* **156** (2009), no. 1, 29–43.

[14] T. Ceccherini-Silberstein, F. Scarabotti and F. Tolli, Representation theory of wreath products of finite groups. *J. Math. Sci.* **156** (2009), no. 1, 44–55.

[15] T. Ceccherini-Silberstein, F. Scarabotti and F. Tolli, *Representation Theory of the Symmetric Groups: the Okounkov–Vershik Approach, Character Formulas, and Partition Algebras*. Cambridge Studies in Advanced Mathematics, vol. 121, Cambridge University Press, 2010.

[16] A.H. Clifford, Representations induced in an invariant subgroup. *Ann. Math.* (2) **38** (1937), no. 3, 533–550.

[17] C.W. Curtis and I. Reiner, *Representation Theory of Finite Groups and Associative Algebras*. Reprint of the 1962 original. Wiley Classics Library. John Wiley & Sons, 1988.

[18] C.W. Curtis and I. Reiner, *Methods of Representation Theory. With Applications to Finite Groups and Orders*, Vols. I and II. Pure and Applied Mathematics Series, John Wiley & Sons, 1981 and 1987.

[19] P. Delsarte, An algebraic approach to the association schemes of coding theory, *Philips Res. Rep. Suppl.* **10** (1973). Available at: http://users.wpi.edu/~martin/RESEARCH/philips.pdf

[20] P. Diaconis, *Group Representations in Probability and Statistics*. Institute of Mathematical Statistics Lecture Notes—Monograph Series, vol. 11, Institute of Mathematical Statistics, 1988.

[21] P. Diaconis and D. Rockmore, Efficient computation of isotypic projections for the symmetric group, in *Proc. Conf. on* Groups and computation (New Brunswick, NJ, 1991), pp. 87–104, American Mathematical Society, 1993.

[22] P. Diaconis and M. Shahshahani, Time to reach stationarity in the Bernoulli–Laplace diffusion model. *SIAM J. Math. Anal.* **18** (1987), 208–218.

[23] J.D. Dixon and B. Mortimer, *Permutation groups*. Graduate Texts in Mathematics, vol. 163, Springer-Verlag, 1996.

[24] L. Dornhoff, *Group Representation theory. Part A: Ordinary Representation Theory*. Pure and Applied Mathematics Series, 7, Marcel Dekker, 1971.

[25] C.F. Dunkl, A Krawtchouk polynomial addition theorem and wreath products of symmetric groups. *Indiana Univ. Math. J.* **25** (1976), 335–358.

[26] C.F. Dunkl, An addition theorem for Hahn polynomials: the spherical functions. *SIAM J. Math. Anal.* **9** (1978), 627–637.

[27] C.F. Dunkl, Orthogonal functions on some permutation groups. in *Proc. Symp. Pure Math.* vol. 34, pp. 129–147, American Mathematical Society, 1979.

[28] A. Figà-Talamanca, An application of Gelfand pairs to a problem of diffusion in compact ultrametric spaces, in *Topics in Probability and Lie Groups: Boundary Theory*, CRM Proc. Lecture Notes vol. 28, American Mathematical Society, 2001.

[29] W. Fulton and J. Harris, *Representation Theory. A First Course*. Graduate Texts in Mathematics, vol. 29, Springer-Verlag, 1991.

[30] P.X. Gallagher, Group characters and normal Hall subgroups. *Nagoya Math. J.* **21** (1962), 223–230.

[31] L. Geissinger and D. Kinch, Representations of the hyperoctahedral group, *J. Algebra* **53** (1978), 1–20.

[32] I.M. Gelfand, Spherical functions on symmetric Riemannian spaces, *Dokl. Akad. Nauk. SSSR* **70** (1959), 5–8 (*Collected papers*, Vol. II, Springer (1988), 31–35).

[33] R.I. Grigorchuk, Just infinite branch groups, in *New Horizons in pro-p Groups*, Progr. Math., vol. 184, pp. 121–179, Birkhäuser, 2000.

[34] L.C. Grove, *Groups and Characters*. Pure and Applied Mathematics John Wiley & Sons, Inc., 1997.

[35] B. Huppert, *Character Theory of Finite Groups*. De Gruyter Expositions in Mathematics, vol. 25, Walter de Gruyter, 1998.

[36] I.M. Isaacs, *Character Theory of Finite Groups*. Corrected reprint of the 1976 original (Academic Press, New York). Dover Publications, 1994.

[37] G.D. James, *The Representation Theory of the Symmetric Groups*. Springer Lecture Notes, vol. 682, Springer-Verlag, 1978.

[38] G.D. James and A. Kerber, *The Representation Theory of the Symmetric Group*. Encyclopedia of Mathematics and its Applications, vol. 16, Addison-Wesley, 1981.

[39] L. Kaloujnine, Sur les p-groupes de Sylow du groupe symétrique du degré p^m. *C. R. Acad. Sci. Paris* **221** (1945), 222–224.

[40] L. Kaloujnine, Sur les p-groupes de Sylow du groupe symétrique du degré p^m (Suite centrale ascendante et descendante). *C. R. Acad. Sci. Paris* **223** (1946), 703–705.

[41] L. Kaloujnine, Sur les p-groupes de Sylow du groupe symétrique du degré p^m (Sous-groupes caractéristiques, sous-groupes parallélotopiques), *C. R. Acad. Sci. Paris* **224** (1947), 253–255.

[42] L. Kaloujnine, La structure des p-groupes de Sylow des groupes symétriques finis. *Ann. Sci. École Norm. Sup.* (3) **65** (1948), 239–276.

[43] A. Kerber, *Representations of Permutation Groups. I.* Lecture Notes in Mathematics, vol. 240, Springer-Verlag, 1971.

[44] A. Kerber, *Applied Finite Group Actions*. Second edition. Algorithms and Combinatorics Series, vol. 19, Springer-Verlag, 1999.

[45] A. Kerber and J. Tappe, On permutation characters of wreath products. *Discrete Math.* **15** (1976), no. 2, 151–161.

[46] M. Krasner and L. Kaloujnine, Produit complet des groupes de permutations et problème d'extension de groupes I. *Acta Sci. Math. Szeged* **13** (1950), 208–230.

[47] M. Krasner and L. Kaloujnine, Produit complet des groupes de permutations et problème d'extension de groupes II. *Acta Sci. Math. Szeged* **14** (1951), 39–66.

[48] M. Krasner and L. Kaloujnine, Produit complet des groupes de permutations et problème d'extension de groupes III. *Acta Sci. Math. Szeged* **14** (1951), 69–82.

[49] S. Lang, $SL_2(R)$. Reprint of the 1975 edition. Graduate Texts in Mathematics, vol. 105. Springer-Verlag, 1985.

[50] S. Lang, *Algebra*. Revised third edition. Graduate Texts in Mathematics, vol. 211, Springer-Verlag, 2002.

[51] G. Letac, Problèmes classiques de probabilité sur un couple de Gelfand, in *"Analytical Problems in Probability"*, Lecture Notes in Mathematics, vol. 861, pp. 93–120, Springer Verlag, 1981.

[52] G. Letac, Les fonctions spheriques d'un couple de Gelfand symetrique et les chaînes de Markov. *Advances Appl. Prob.* **14** (1982), 272–294.

[53] W.C.W. Li, *Number Theory with Applications*. Series on University Mathematics, vol. 7, World Scientific, 1996.

[54] J.H. van Lint and R.M. Wilson, *A Course in Combinatorics*. Second edition. Cambridge University Press, 2001.

[55] G.W. Mackey, Unitary representations of group extensions. I. *Acta Math.* **99** (1958), 265–311.

[56] G.W. Mackey, *Unitary Group Representations in Physics, Probability, and Number Theory.* Second edition. Advanced Book Classics, Addison-Wesley, 1989.

[57] H. Mizukawa, Zonal spherical functions on the complex reflection groups and $(n + 1, m + 1)$-hypergeometric functions, *Adv. Math* **184** (2004), 1–17.

[58] H. Mizukawa and H. Tanaka, $(n + 1, m + 1)$-hypergeometric functions associated to character algebras. *Proc. Amer. Math. Soc.* **132** (2004), no. 9, 2613–2618.

[59] A. Okounkov and A.M. Vershik, A new approach to representation theory of symmetric groups. *Selecta Math. (N.S.)* **2** (1996), no. 4, 581–605.

[60] J.J. Rotman, *An Introduction to the Theory of Groups.* Fourth edition. Graduate Texts in Mathematics, vol. 148, Springer-Verlag, 1995.

[61] B.E. Sagan, *The Symmetric Group. Representations, Combinatorial Algorithms, and Symmetric Functions.* Second edition. Graduate Texts in Mathematics, 203, Springer-Verlag, 2001.

[62] J. Saxl, On multiplicity-free permutation representations, in *Finite Geometries and Designs*, London Mathematical Society Lecture Notes Series, vol. 48, p. 337–353, Cambridge University Press, 1981.

[63] F. Scarabotti and F. Tolli, Harmonic analysis of finite lamplighter random walks. *J. Dyn. Control Syst.* **14** (2008), no. 2, 251–282.

[64] F. Scarabotti and F. Tolli, Harmonic analysis on a finite homogeneous space. *Proc. Lond. Math. Soc.* (3) **100** (2010), no. 2, 348–376.

[65] F. Scarabotti and F. Tolli, Harmonic analysis on a finite homogeneous space II: the Gelfand–Tsetlin decomposition. *Forum Math.* **22** (2010), no. 5, 879–911.

[66] C.H. Schoolfield, A signed generalization of the Bernoulli–Laplace diffusion model. *J. Theoret. Probab.* **15** (2002), no. 1, 97–127.

[67] J.P. Serre, *Linear representations of finite groups.* Graduate Texts in Mathematics, vol. 42, Springer-Verlag, 1977.

[68] B. Simon, *Representations of Finite and Compact Groups.* American Mathematical Society, 1996.

[69] R.P. Stanley, *Enumerative Combinatorics,* vol. 1. Cambridge University Press, 1997.

[70] D. Stanton, Orthogonal Polynomials and Chevalley Groups, in *Special Functions: Group Theoretical Aspects and Applications* (R. Askey *et al.*, Eds.) pp. 87–128, Dordrecht, 1984.

[71] D. Stanton, Harmonics on Posets. *J. Comb. Theory Ser. A* **40** (1985), 136–149.

[72] D. Stanton, An introduction to group representations and orthogonal polynomials, in *Orthogonal Polynomials* (P. Nevai, Ed.), pp. 419–433, Kluwer Academic, 1990.

[73] S. Sternberg, *Group Theory and Physics.* Cambridge University Press, 1994.

[74] A. Terras, *Fourier Analysis on Finite Groups and Applications.* London Mathematical Society Student Texts, vol. 43, Cambridge University Press, 1999.

[75] H. Wielandt, *Finite Permutation Groups.* Academic Press, 1964.

[76] E. Wigner, On unitary representations of the inhomogeneous Lorentz group. *Ann. Math.* (2) **40** (1939), no. 1, 149–204.

Index